3D Printing

Recent Titles in the Libraries Unlimited Tech Tools for Learning Series

The Networked Library: A Guide for the Educational Use of Social Networking Sites
Melissa A. Purcell

Bookmarking: Beyond the Basics
Alicia E. Vandenbroek

School Library Infographics: How to Create Them, Why to Use Them
Peggy Milam Creighton

3D Printing: A Powerful New Curriculum Tool for Your School Library
Lesley M. Cano

3D Printing

A Powerful New Curriculum Tool for Your School Library

Lesley M. Cano

Tech Tools for Learning

Judi Repman, Series Editor

An Imprint of ABC-CLIO, LLC
Santa Barbara, California • Denver, Colorado

Library of Congress Cataloging-in-Publication Data

Cano, Lesley M.
3D printing : a powerful new curriculum tool for your school library / Lesley M. Cano.
pages cm.—(Tech tools for learning)
Includes bibliographical references and index.
ISBN 978-1-61069-977-8 (paperback) — ISBN 978-1-61069-978-5 (ebook) 1. School libraries—Activity programs. 2. Three-dimensional printing. 3. School librarian participation in curriculum planning. I. Title. II. Title: Three D printing.
Z675.S3C194 2015
027.8—dc23 2015012916

ISBN: 978-1-61069-977-8
EISBN: 978-1-61069-978-5

19 18 17 16 15 1 2 3 4 5

This book is also available on the World Wide Web as an eBook. Visit www.abc-clio.com for details.

Libraries Unlimited
An Imprint of ABC-CLIO, LLC

ABC-CLIO, LLC
130 Cremona Drive, P.O. Box 1911
Santa Barbara, California 93116-1911

This book is printed on acid-free paper ∞

Manufactured in the United States of America

To Mel, for her unflagging optimism and enthusiasm.

To Bob, Olivia and BJ, and Paula, for their support and encouragement.

And to Noah, Olivia, and Gabrielle, just for being.

Contents

Acknowledgments

I would like to thank a few people who have made my foray into 3D printing a little easier and a lot more successful than it would have been without them. First and foremost, I would like to thank Matt Brown, and Selena Ozuna, past and present administrators of Remynse Elementary in Arlington, Texas. They have both been champions for my school library program and are two of my strongest supporters.

When I first approached Matt with the idea of creating a makerspace in our school library, he didn't think twice before responding with a resounding yes. He enabled me to create a library that truly has transformed the way our students learn. He not only supports me as a librarian, he provides me with the challenge I need to reach beyond my comfort zone as a leader and encourages me to think outside of the box. I appreciate that both Matt and Selena recognize the power of a strong school library program and the difference school librarians can make.

Thanks also to Julie Moore, Library Media Services Coordinator extraordinaire. Above all, a school library program needs a strong leader who is passionate about the power of school libraries and who puts the good of students above all else, and that's exactly what Julie does. She advocates relentlessly for school libraries and her passion keeps me moving forward to provide 21st-century learning opportunities for students every day. Her indefatigable work and leadership are inspirational.

Much thanks to the faculty and staff at Remynse Elementary for their words of encouragement. Writing a book can be an arduous process, and knowing they were behind me made all the difference. Much gratitude and appreciation goes to Katy O'Connell, for her extra helping of assistance, support, and encouragement.

Most of all, I need to thank the students at Remynse Elementary. The day we took our first 3D printer out of the box, a student told me with tears in his eyes that he never thought he would get to see a 3D printer in real life, let alone get to use one. Now, that same student proudly displays several 3D printed objects of his own design in the library. This student, like each of my students, has displayed enthusiasm and determination in learning new things, and patience with me as we have learned together.

Introduction

Do you get excited about discovering new ways your library can impact student achievement and transform how your students learn? What if it involved technology so advanced that it not only is currently being used in a variety of fields to change the lives of people around the world but also is readily available to your K–12 students? Welcome to the world of 3D modeling and printing!

3D modeling and printing can open doors for critical thinking, problem solving, and creativity in students and can be a catalyst for dynamic and engaging educational experiences. 3D modeling and printing has given me the opportunity to witness a transformation in student thinking. Students, including those who had become disenchanted with school, are now excited about learning and love coming to the library!

Two years ago, 3D modeling and printing were not on my radar when I was thinking of how I could better reach students through the power of the school library. Although I love technology and am fascinated by how technology engages students, I have never considered myself to be particularly proficient at figuring technology out, and frankly, I was a little unsure if I would be able to learn 3D modeling and printing so I could teach it to my students. The good news is, I didn't have to.

You don't have to be adept at 3D modeling and printing in order to get it started in your school library. I learned 3D modeling and printing right along with my students. They loved seeing me learning right alongside them and enjoyed it immensely when they could teach me a thing or two.

I became interested in 3D printing because I am always looking for ways to keep the library relevant. I teach at a Title I school where almost 90 percent of students are low income and English is a second language to many of our students. I strongly feel it is my duty to expose students to the newest technologies so they will have skills to compete for college and careers when they are ready. This encourages me to continue to think of what and how I can reach students in a new and engaging way.

When I first read about 3D printing, I had a hard time imagining exactly what it was. I read about a high school utilizing 3D printing as part of an engineering program. My first thought was, "Wow! That's amazing that high school students are doing this!" My second thought was, "Why can't my elementary students do this?" After a lot of research and talking with many people about using 3D printing in schools, I gathered my notes and went in to speak with my principal. I am lucky to work for a principal who not only loves technology, but also appreciates how technology can transform learning and teaching. He approved a grant I wrote through my district's Instructional Technology department and through this grant I received robotics kits and a 3D printer for my library's makerspace.

When I first unpacked our 3D printer, I honestly wasn't quite sure what to do with it. As with most new technology, I had to read the directions step-by-step but ultimately got it set up with a quick call to the company's support team. There were several ready-to-print models already on the USB drive that came with the printer, so I printed one of these sample objects. We had a lot of fun with it, finding more models of cool things to print, such as Batman and chess pieces. I began to think of how 3D modeling and printing could be utilized across the curriculum and I invited teachers in my building to do the same. Together, we began to discover ways to integrate 3D modeling and printing into what was going on in the classroom. Our elementary students learned to design 3D models using Tinkercad and they went from printing existing models to designing and printing their own.

It is inspiring to see first-grade students designing 3D models and their excitement when they held the objects they had designed in their hands. When students began to design their own 3D models, this was a huge step and made the process much more rewarding and beneficial for students.

Something I discovered along the way is that learning 3D modeling and printing is a process. It takes a little time to learn. As with learning most new things, some students will pick up the concept very quickly and others will need a little more support, encouragement, and time to master it. Some students will begin to print complex and detailed objects early on in the process, while others will stick to more basic designs. Sometimes, objects will print that don't look so great. All of these are ok because it's all part of the process. Let students go through the process at their own pace but also don't be afraid to push them a little. Make it fun by providing incentives through contests and other exciting events. Students will gravitate to 3D modeling and printing and will love learning it, even though they may struggle sometimes with designing an object that prints out exactly like they envision it.

This book was born out of my desire to not just have a 3D printer for student use, but to have it used as a part of the curriculum to support student learning. I wanted to harness the excitement I saw that students had for 3D modeling and printing and use it to build interest in what was going on in the classroom and the library. This book is intended for school librarians who are brand new to 3D modeling and printing. It is intentionally not an overly technical book about 3D modeling and printing because although technical details are nice to know, in my experience they aren't really necessary for students to begin modeling and printing 3D objects.

The first chapter covers the basics of 3D printing in a way that is easy to understand to give librarians an overview of what 3D modeling and printing are and the basics of what supplies are needed to get started. Chapter 2 discusses the impact that 3D modeling and printing can have on students and would be great to use as a resource to write a grant to get a 3D printer or to state your case for purchasing one. There are several chapters on 3D modeling programs that provide basic how-to's to get students started. I have intentionally included a wide variety of programs to help you find a good one for you and your students. The variety will ensure students can each find a program they are comfortable with and enjoy using.

Individual chapters focusing on math, science, language arts, social studies, and fine arts will give ideas for how 3D modeling and printing can be integrated into the curriculum. Again, these are meant to be a starting point. Although the lesson ideas are listed by grade level, they can certainly be adjusted to fit a wide variety of grade levels, interests, and projects. The last chapter provides practical tips and tricks for using 3D modeling and printing in your library.

It is important to note that while several brands and models of 3D modeling programs and printers are mentioned in this book, these by no means constitute an all-inclusive list. 3D modeling and printing is progressing at such a fast rate that new models appear quickly. There are many software programs and many types, brands, and models of 3D printers to choose from, but most have similar features. My hope is that you can use this book as a springboard to introduce 3D modeling and printing to students in your library. I am a school librarian just like you. I know the challenges, and I know that most things don't work the way they are presented in books. Don't be afraid to start, even if you have to start small. Your students will benefit and thrive for your efforts. Be creative. Take risks. Challenge students. Challenge yourself.

1

Basics of 3D Modeling and Printing

Some educators think that 3D printing is just the latest in a long line of technology fads. Others believe that while 3D printing is a valuable tool in manufacturing and industry, it has little value to students in educational settings. Many educators don't see the value of 3D printing to education, curriculum, and ultimately, to children. The fact is that a 3D printer can not only be a fun and engaging addition to a school and, more specifically, a school library, it can also be integrated into the curriculum in a way that fuels student curiosity and creativity and also ignites students' passion for learning.

Educators have heard in the past few years that they must prepare students for jobs that haven't even been created yet due to the exponential growth of technology. What does this mean for school librarians? Librarians have to figure out how to teach skills that will enable students to be successful in these jobs that don't yet exist. Students need to learn a skill set that can be applied to multiple situations and jobs in the future. Educators may be unable to prepare students for *specific* jobs of the future, but they definitely can teach students the 21st-century skills they need to succeed in these jobs.

What skills do students need to carry them far into the future? Creativity, problem solving skills, critical thinking skills, and the ability to collaborate and communicate on a global level are just a few. Although these skills are of utmost importance, these types of skills can also be the most challenging for educators to teach. These are often viewed as more abstract than math or reading skills by educators and thus educators think they are difficult to teach. Many educators also question their own ability to teach these skills when so much of what they do is focused on preparing students for high-stakes testing.

So how can an educator build 21st-century skills in students? Providing students with innovative technology tools is an effective way to give students opportunities to gain and practice these skills. 3D modeling and printing are ideal tools to teach these 21st-century skills that will also engage and inspire students.

In their 2013 report "The Technology Outlook for STEM+ Education 2013–2018: An NMC Horizon Project Sector Analysis," experts from New Media Consortium (NMC), the Centro Superior para la Enseñanza Virtual (CSEV), the Departamento de Ingeniería Eléctrica, Electrónica y de Control at the Universidad Nacional de Educación a Distancia (UNED), and the Institute of Electrical and Electronics Engineers Education Society (IEEE) conducted research on emerging technologies in education. These experts named 3D printing as one of the emerging technologies that will most impact STEM education. This study found that 3D printing allows for more authentic exploration of objects that may not be readily available to educational institutions, such as animal anatomies and toxic materials. The exploration of 3D printing, from design to production, as well as demonstrations and participatory access, can open up new possibilities for learning activities. Typically, students are not allowed to handle fragile objects like fossils and artifacts; 3D printing shows promise as a rapid prototyping and production tool, providing users with the ability to touch, hold, and even take home an accurate model (Johnson et al. 2013).

Research findings are clear. 3D modeling and printing address the sort of STEM skills that many educators and policymakers consider most important to productivity in the 21st century.

A Very Brief History of 3D Printing

It seems as though 3D printing is a fairly new technology because of all the media surrounding it, but 3D printing has been around for quite a while, albeit not in the realm of education. In his article, *Manufacturing the Future: How 3D Printing Went from Pipe Dream to Your Desktop*, Brian Heater (2014) describes how stereolithography was created when Chuck Hull discovered how to add another dimension to lithography, an 18th-century printing technique. This was the process of stereolithography, which used lasers and liquid resin to create 3D objects.

3D printing has been added to more elementary and secondary schools in the last few years and has surged in numbers recently. Some secondary schools began to acquire 3D printers as part of engineering courses, but this was the exception rather than the rule. Although 3D printers have primarily been used as part of a makerspace or engineering lab setting in schools, librarians have begun to recognize the potential of 3D modeling and printing in school libraries. Educators are now beginning to see the significance of adding 3D printers to a school setting due to the media surrounding 3D printing. It's rare when there is a day without news of how 3D-printed prosthetics, shelters, or other objects have made someone's life better.

Additive Technology Manufacturing

3D printing is a type of additive technology manufacturing, which means that instead of starting with a solid block of material and cutting it down to a shape, an object is built by adding one layer of material at a time. 3D printing is only one type of additive technology manufacturing. Other types include inkjets, aerosol jetting, electron beam melting, and laser sintering.

The major benefit of additive technology manufacturing is that objects are created without the use of molds or dies. Additive technology manufacturing allows for objects to be created on demand, anytime the object is needed. Building objects with additive technology manufacturing keep the manufacturing cost down, as design changes can be made quickly and easily with little or no additional cost.

How Does 3D Printing Work?

The 3D printing process can be divided into two stages: modeling and printing. The modeling stage is where the user creates a digital model of the object using software compatible with the particular printer being used. The model is then uploaded to the printer and printed in thin layers. In most consumer 3D printers, the material used to create the object is a type of plastic. Most 3D printers print by using the concept of Cartesian robotics. The printer moves linearly along the *X*, *Y*, and *Z* axes based on a model created in a 3D modeling software program. The *X* axis specifies the width of the model and the *Y* axis specifies the length. The *Z* axis is what gives the 3D model height.

What Is 3D Modeling?

3D modeling is the creation of a digital representation of an object. 3D models have length, width, and depth. 3D modeling allows the creator to draw, build, and manipulate the digital object on the computer. Users can scale the model by increasing or decreasing the size, rotate the digital model, or change the model in any way. After a model is created, it is saved in the .STL file format. The .STL file format is the standard format used for 3D printing and contains a list of the *X*, *Y*, and *Z* coordinates of the vertices for the object being printed. The printing software evaluates data from the model and then converts that model into layers, or slices, which changes the file from .STL to G-Code.

What Is the Difference Between .STL and G-Code . . . and Do I Need to Know This?

A simplified explanation of the difference between .STL files and G-Code is that .STL files are the computer files and the G-Code is the file that is actually printed. .STL files are converted to G Codes by importing the .STL file into the software for the particular 3D printer being used. The software takes the .STL file through the slicing process, which turns the .STL file into the G-Code that can then be printed by the 3D printer by determining characteristics of the object.

During the modeling and design process, the digital representation of the object must meet the exact specifications the user wants the object to have after it is printed. Students are often surprised when the object printed meets the exact specifications of the object designed, including the flaws of the model. Part of the value of the process is that students learn to convert what they see on the screen to an image of what the actual object will look like after printing.

The model has to be uploaded to the 3D printer, and this can be done using a USB drive or SD card. Wi-Fi connection can be used if the 3D printer is capable of connecting in this manner. This allows the model to be sent straight to the printer. Caution should be used when allowing students to use Wi-Fi to send 3D models straight from the modeling program to the printer; some of the models may not be printable due to problems in the design, and this will result in wasting of filament.

The good news is that librarians don't really need to know details of how .STL and G-Code files work. The modeling programs used really do this part for you, but it *is* helpful to know what .STL and G-Code files are so you can teach students to use the correct terminology and recognize what these files are when they see them.

The 3D Printing Process

The printing software slices the object into thin layers, and this enables the printer to lay down the material (usually a type of plastic in consumer 3D printers) in layers to build the final object. Melted plastic is extruded into these thin layers according to the *X*, *Y*, and *Z* coordinates set by the design process. The layers begin at the bottom of the object and build the object from the bottom to top, layer by layer, until the object is complete. The thickness of each layer is determined by the diameter of the plastic filament used and the size of the extruder on the printer. The thinner the layers of plastic, the stronger and more detailed the object will be when printed. Although thinly printed layers help to ensure a strong and detailed object, the downside is that it takes longer to print, so patience can truly be a virtue when it comes to 3D printing large objects.

Parts of a 3D Printer

Although 3D printers vary according to brand, size, and type of material used, they all have basically the same parts. While it is not imperative that students learn the parts of the printer, it is helpful if they are exposed to and use the correct terminology.

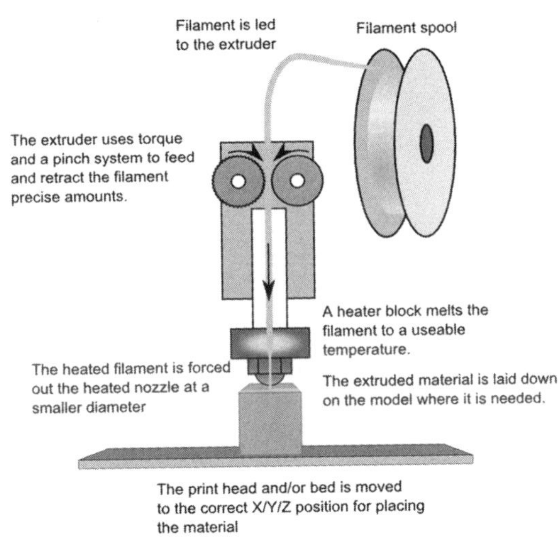

How 3D Printers Work (Licensed under Creative Commons Attribution license 3.0. Retrieved from http://www.thingiverse.com/thing:29432)

Extruder

One of the most important parts of a 3D printer is the extruder. The extruder is the small metal piece of the printer where melted plastic comes out, or extrudes. The extruder becomes hot, so extreme care should be taken to supervise students when they are observing the printing of an object.

Extruder

Although the extruder is usually insulated in some manner to help prevent burns, it is imperative students are taught to not touch any part of the printer, whether during a print or not, without permission. This teaches students to respect the printer as a heat source and helps keep everyone safe.

Most consumer 3D printers have one extruder, which allows for one material and color to print at a time, while other printers have two or more extruders, allowing for multiple colors of filament to print an object at the same time. Generally, the more extruders a printer has, the more expensive the machine is.

Print Bed

The print bed is the flat surface the melted filament is laid on. A print bed comes with the printer when purchased and can be made of glass, acrylic, metal, or ceramic and can be heated or unheated, depending on the 3D printer. Usually some type of stabilizing material has to be added to the print bed to ensure the object stays secure as it prints. Sometimes, stabilizing material such as glue is included with the printer along with information on how to purchase more. Less expensive and easily found stabilizing materials can also be used:

Blue painter's tape – This tape adheres to the print bed but is easy to remove when it needs to be replaced. The tape needs to be replaced when it gets worn, which is usually every few prints, but it is relatively inexpensive so replacing it often is not a problem.

Hairspray – Extra-super-hold hairspray can work well on some print beds and not so well on others, so experimentation will be needed if this stabilizer is chosen.

Glue stick (such as Elmer's All Purpose glue stick) – This works well, especially on acrylic print beds.

If using a type of glue, the object may have to be gently pried from the print plate after printing with a scraper possessing a thin edge. A painter's scraper or metal label peeler work well for this because their thin edges slide between the print bed and the printed object easily. It can help to soak the print bed (with the object attached) in warm water so the object releases cleanly and without breaking. If using glue, the print bed will need to be cleaned and properly prepped before each new print.

There are companies now experimenting with creating 3D printer beds that do not require the use of glue or tape as a stabilizer. One of these is the GeckoTek print bed. The makers of the GeckoTek print bed have created a print bed with a permanent coating that stabilizes an object as it is printed. The object also can be easily removed from the GeckoTek material after printing. The GeckoTek print bed would replace the print bed packaged with the 3D printer.

Leveling the Print Bed

To ensure high-quality printed objects, the print bed needs to be leveled on a regular basis. Leveling the print bed is one of the most important tasks to make sure the extruder stays the same distance away from every point on the print bed, thereby ensuring the filament sticks evenly. Ensuring the first layer of material is consistently adhered to the print bed will help guarantee a good-quality first layer, which leads to a good-quality print overall.

The print bed should be leveled before every new print. If the 3D printer has not been in use for a while, the print bed will need to be leveled before being used again. If you have traveled with the 3D printer or if it has been moved, the print bed will probably need to be leveled. The general rule for leveling is: If in doubt, go ahead and level it out!

If the extruder nozzle is too close to the print bed, the filament will not extrude smoothly and consistently and may cause the extruder to jam. If the extruder nozzle is too far from the print bed, the filament may not stick to the print bed and it won't be building the desired object. Instead, filament will extrude but not be laid down in a pattern, causing a mess on the plate, and in cases such as this, the extruder nozzle can clog. If this happens, the 3D printer may have to be disassembled and cleaned before it can be used again. This is one of the reasons it is important the 3D printer is not left unattended during printing.

How to Level the Print Bed

This is a job that older students can certainly be taught to perform before each print job. The only tool needed to level the print bed is a piece of paper. Some 3D printers have small knobs that the user manually adjusts, and other 3D printers have a leveling process on the built-in menu for leveling the print bed. If the latter is the case, the user just needs to read and follow the instructions on the screen. To level the print bed, use the buttons on the menu or the mounted screws to increase the gap between the extruder nozzle and print bed. After increasing the gap, insert a piece of regular printer paper between the extruder nozzle and the print bed.

Use the menu button to gradually decrease the gap between the extruder nozzle and the print bed until the paper passes between them with a slight amount of friction. If the paper doesn't fit between the extruder nozzle and print bed, the gap needs to be increased. If the paper passes between the nozzle and bed easily with no friction at all, the gap needs to be decreased. Leveling the print bed sounds more difficult than it really is. Leveling is extremely easy to do and will become second nature after doing it a few times. Although the general guidelines for leveling given here will work for most printers, finding the guidelines of leveling the print bed of the specific printer being used is the best course of action. This information can be found in the information that came with the printer or online at the manufacturer's website. Users can also search YouTube for videos that other users have posted showing how to level the bed on a particular brand and model of printer. Remember to never try to level the print bed by directly touching the extruder nozzle or print bed; let the machine do it for you.

Removable Print Beds

Many 3D printers have removable print beds. Removing the print bed makes it much easier to release the printed object from the bed. If the print bed is removable, never try to release the printed object while the plate is still attached to the printer. Doing so could cause damage to the printer. Remove the print bed, release the object, and the bed can be prepared and inserted again to begin a new print. To speed up printing, a second print bed could be purchased so that as students are releasing an object from one print bed, the other could be inserted and a new print could begin.

Printing with Rafts

One way to help ensure a strong print and to try to overcome a print bed that is not perfectly level is to print the object with a raft. In 3D printing terms, a raft is a woven layer of sparse layers of filament that provide a base for the object being printed. The object is printed on top of the raft, which provides the object with a stable base. After the object is printed, the raft is detached from the object.

Choosing a 3D Printer

There are many attributes to consider when selecting a printer for a school library. Begin by reflecting on the types of objects students will ultimately be printing. For younger students who will be primarily printing simple objects, a smaller, less-expensive printer should be sufficient. Otherwise, it might be wise to consider getting a larger printer that can better print objects with sharp details. Older students may begin with smaller, less-complex objects, but once they get the hang of 3D printing and get excited about the possibilities, they will quickly begin modeling and printing larger, more-complex objects.

The most basic specifications that should be considered when deciding on which 3D printer to purchase are print area, print volume, print speed, and print surface.

Print Area and Volume

Refers to the largest size at which an object can be printed. For most school-related projects, an average print area of at least 5 in. × 5 in. × 5 in. is adequate, but again, a larger printer will allow students to grow into it and not be limited by print area.

Print Speed

The speed at which an object is printed. Although a faster print speed might be seen as beneficial, for most school settings, it isn't a big issue. If students begin a printing project that takes a few hours, they can check back at lunch, after school, or the next day. A print speed of at least 20 mm per second should be more than adequate.

Print Surface

The type of surface the object is printed on and whether it is a heated surface or not. This refers to the print bed or plate mentioned earlier.

There are many brands of 3D printers. Below is a chart of popular brands and models of consumer 3D printers.

Manufacturer	*Model*	*Approximate Cost*	*Type of Filament Used*	*Footprint*	*Print Area*	*Print Bed*
Cubify	Cube	$999	PLA and ABS	13.2″ x 13.5″ x 9.5"	6 x 6 x 6	Not heated
Lulzbot	Mini	$1350	PLA and ABS (plus many more materials)	20″ x 20″ x 27″	6″ x 6″ x 6.2″	Heated
Makerbot	Replicator	$2899	PLA	19.1″ x 12.8″ x 14.7″	9.9″ x 7.8″ x 5.9″	Not heated
Printrbot	Simple Maker (kit)	$349	PLA only	15″ x 11 1/2″ x 10″	4 x 4 x 4	Not heated
Ultimaker	Ultimaker 2	$2500	PLA and ABS	14″ x 13″ x 15″	9″ x 9″ x 8″	Heated

RepRap

For the more adventurous, RepRap is a creative way to acquire a 3D printer. RepRap, which is short for "replicating rapid prototyper," is a self-replicating 3D printer that was originally created by Dr. Adrian Bowyer in 2005. The term "self-replicating" means that the printer prints itself. RepRap is a 3D printer that prints plastic parts, and can print a kit of itself that can be assembled to create a new RepRap. Someone with a RepRap, and plenty of time and patience, could 3D print another RepRap. RepRap.org is an online community to help those that want to build their own 3D printers. RepRaps are definitely for the makers who not only enjoy 3D printing, but would also enjoy the challenge of building their own printer. If this is something students are interested in, the RepRap.org website is a great place to begin for resources on getting started.

Filament

Along with the initial cost of the 3D printer, the primary ongoing expense of 3D printing is the filament that the printer uses to build the objects. Before purchasing a printer, users should check to see if the manufacturer requires that a proprietary brand of filament be used, or if a more generic brand, which costs less, can be used. Most printers accept filament that can be purchased from a variety of sources, but there are a few printers that require a particular brand of filament, which can add to the cost. For example, Cubify printers require that only Cubify filament be used, and other filament brands cannot be used. The Cube filament comes encased in a cartridge that fits only that particular printer.

PLA filament

Although some 3D printers can print in specialty materials, the most commonly used filament in consumer 3D printers is made from plastic. Two types of plastic filament are used in 3D printing: ABS (acrylonitrile butadiene styrene) and PLA (polylactic acid). Most consumer 3D printers can print using either ABS or PLA filament. Some 3D printers can use both ABS and PLA filament, whereas others are made to work with only one of them. ABS and PLA plastics are thermoplastics, which means they can be heated to the point of being malleable and then return to a hardened state as the plastic cools. Filament is made in diameters of 1.75 mm and 3 mm. Because different 3D printers use different diameters of filament, make sure that the proper diameter of filament is purchased and used.

ABS filament is petroleum based and comprised of acrylonitrile, butadiene, and styrene, which some research has found to be toxic, especially when heated. Although there is not a lot of research on the dangers of ABS fumes, the studies conducted so far indicate there is need for caution. A study conducted by Stephens et al. found that ABS filament gave off almost twice as many ultrafine particle emissions than PLA filament. Dangers from ABS fumes were studied in a 1994 project by Schaper, Thompson, and Detwiler-Okabayashi, who found that the fumes resulted in respiratory distress in laboratory mice. Because of this, and because the safety of students is always top priority, it is recommended that ABS filament *not* be used in a school setting, even if the 3D printer is in a well-ventilated area. There are too many unknowns regarding the fumes given off when ABS is heated during the 3D printing process to take the risk of using it around children.

PLA filament is made from plant-based materials. PLA can be made from sugar cane, beet cane, corn, or other vegetable waste and is biodegradable. Although biodegradable, PLA won't biodegrade in the typical compost bin, as higher temperatures are needed for it to break down. This type of filament is considered safe to use to print objects that will come into contact with food. PLA may have a strong odor when heated, but research so far has shown it to be safe when heated. Because PLA is made from natural products, it can have a smell when heated, but students usually compare it to the smell of warm syrup. Ensuring that the 3D printer is in a well-ventilated area or is ventilated to the outside is wise regardless of the type of filament used. An additional benefit of PLA is that it resists warping, even when used to print larger objects.

Nylon trimmer line (the type of line used in weed whackers) is being used by some 3D print enthusiasts, but there has been no research yet on the dangers of fumes given off by the trimmer line when it is heated. Therefore, it is not recommended that trimmer line be used as a filament in a school setting.

Other materials used for printing with more advanced 3D printers are foam, clay, polycarbonate, metal, chocolate, and even human stem cells. Almost every day, there are numerous stories of new ways 3D printers are being used to improve the lives of people around the world. The boundaries of 3D printing are being pushed, resulting in advances being made quickly in how printers and materials are used for printing.

Tips for Filament Storage

Unopened plastic filament can be stored for an extended period of time. Once a roll of filament has been opened, it can be affected by humidity. If a roll is partially used and needs to be stored, it is helpful to store the roll in an airtight container. A desiccant placed in the container along with the filament roll will provide extra protection against humidity. Desiccants are small packets, usually of silica, that can be found in packages to prevent items from moisture damage. These desiccant packets can easily be collected and stored by older students. Caution should be used when using desiccant packets in areas where younger children will have access to them. These should be kept out of reach of younger students who might mistake the packets for food or candy.

Things to Consider before Purchasing a 3D Printer

Before purchasing a 3D printer for student use, ask the following questions:

Does the outside of the printer get hot?

What heated parts are exposed and could be touched by students?

Are the heated parts enclosed or insulated?

Is the plate heated?

What type of power supply cord does the machine use and does it match the outlets available?

Does the printer require a certain brand or size of filament?

What is the cost of each filament cartridge? (Most filament cartridges cost around $50, but suppliers usually give a small discount for purchasing in larger quantities.)

How much filament should I order? Check with the filament manufacturer to find the estimation of the number of objects each cartridge will print and estimate how much will be needed.

What color(s) of filament should I order? PLA filament comes in a variety of colors. Consider the types of objects students will be printing. You may want to order basic black and white, or you may want to order basic colors along with more vivid colors that students will enjoy such as neon orange, red, blue, or purple.

How Can I Afford a 3D Printer?

One of the primary challenges when purchasing a 3D printer, especially in schools, is the cost. Although the price of 3D printers has decreased dramatically in the past few years, the high cost continues to prohibit librarians from purchasing them. Even when there is a desire to add a 3D printer to a library, schools have limited funds and have to prioritize when purchasing new technology. Technology in schools is often purchased based on the "best bang for the buck" model—the number of students that will use and benefit from the technology in relation to the price of the item. This often means that items such as 3D printers are placed at the bottom of the purchasing list.

This leaves librarians to seek other means of obtaining money for purchasing 3D printers. One of the most successful ways of finding money for a 3D printer involves grant writing. A good place to start is to research local sources for grant funding such as school district grant foundations. Another source for technology grants is large corporations. Most large companies have a department devoted to community outreach, which includes grants, many of which are just for education. Grants that focus on innovative technology or STEM are perfect avenues for obtaining a 3D printer. Search websites that specialize in listing grants for education.

Website sources for finding grants:

getedfunding.com

eschoolnews.com/funding

grantwrangler.com

A great resource for grant writing, and any other education-related topic, is Livebinders.com. My colleague and fellow school librarian, Alicia Vandenbroek, has an extensive collection of grant resources in the livebinder she created for school librarians and teachers that is consistently one of the top-ten viewed binders on the website. To access this vast collection of sources, search the Livebinder.com website for Binder ID # 131571.

There are many resources, books, and websites to help educators write grants. Taking the time to find a grant that focuses on educational technology and then writing the grant with a focus on innovative ways to use 3D modeling and printing with students will help sell the idea and ultimately, win the grant.

Parent-teacher associations or local community organizations are also possible funding sources for educational technology. Sometimes even parent-teacher associations will help finance part of the cost or help with fundraising opportunities.

Another option for funding the purchase of a 3D printer is to utilize a crowdfunding source. Crowdfunding sites allow people with big ideas but little money to post information regarding their project. Others can go on the site, search through giving opportunities, and donate immediately to a chosen project. Crowdfunding sites allow educators to easily reach many potential donors looking for a project to help fund. The following are some crowdfunding sites specifically for educators:

www.adoptaclassroom.org/

www.incited.org/en

crowdfundedu.com/

www.donorschoose.org/

3D Printing Pens

3D printing pens are a recent addition to the field of 3D printing tools. These 3D printing pens are portable versions of the larger 3D printers. Two of the most popular 3D pens are the 3Doodler and the Lix Pen. The 3Doodler pen was successfully funded on Kickstarter.com and now is sold through the company's website, the3doodler.com. The Lix Pen (lixpen.com) was also funded successfully on Kickstarter.com and is marketed as the smallest 3D printing pen in the world. The 3Doodler is being sold now for around $100, and the Lix Pen will soon be in the preorder stage and will cost $60–$140, depending on which of the two pen options is chosen.

3D printing pens work similarly to 3D printers, but without a print bed. Users draw in the air or on a surface while the filament extrudes to build an object. Although a little trickier to use than a 3D printer, older students will enjoy learning to use it. Users load the pens with strands of ABS or PLA filament and wait a couple of minutes for it to heat up. The pen extrudes the filament as the user moves it to lay the filament to design an object. Drawing on a flat surface or tracing a design is easier, while learning to design vertically from the bottom up is more of a challenge. While probably too complex for younger students to handle, these 3D printing pens may be a viable option for 3D printing for secondary students.

There are more 3D printing pens in the pipeline, including the CreoPop pen, which got its start on the crowdfunding site, Indiegogo.com. The CreoPop pen uses a special ink, called photopolymers, that is sensitive to light and solidifies using an ultraviolet light, which is built into the pen. Because the pen uses photopolymers and UV light instead of melted plastic, there are no heated parts on the pen.

3D Modeling Programs

Before 3D printing can occur, a model has to be created. There are some websites that provide digital models that have already been designed by others. These 3D models can be downloaded and printed immediately as is, or changes can be made to the objects before printing. Thingiverse.com is a site that has many models available and is fairly easy for students to use. Allowing students to experiment with printing models that have already been created is a good way to introduce the concept of 3D modeling and printing to students. After students understand the concept of 3D modeling and printing using models already created, they can move toward building original models they design themselves and print.

There are three stages in utilizing the 3D modeling and printing process with students.

1. Students find and print 3D models created by others.
2. Students find, modify, and print 3D models created by others.
3. Students use 3D modeling programs to design and print their own models.

Stage 1: Students find and print models created by others.

Stage 2: Students find, modify, and print models created by others.

There are many websites that provide access to 3D models already created. Models on these sites range from simple to complex. Users can search these sites and find an object to print. The files for the object can be modified or downloaded as is and sent to the 3D printer. This can be done via USB, SD card, or WiFi connection.

Some of the websites where students can find 3D models already created are:

www.Thingiverse.com

www.Repables.com

www.Youmagine.com

www.Cubify.com

www.myminifactory.com

www.autodesk123D.com

http://www.instructables.com/group/123d

Stage 3: Students use 3D modeling programs to design and print their own models.

There are many design programs and apps available for 3D modeling, but fewer that are kid-friendly and easy to use. Some of the modeling programs that are fairly easy for students to learn include Tinkercad, 123D Design, 3D Tin, and SketchUp. These programs are covered in detail in chapters 4–7. There are many more-complex 3D modeling programs, and these may be utilized if students outgrow the less-complex programs. Matching the modeling program to the skill level of students will help keep students engaged and excited about the 3D modeling and printing process. There are also student-friendly apps for 3D modeling on tablets. These apps, including Blokify and 123D Design, are covered in detail in Chapter 8.

How 3D Printing Is Being Used in Industries across the Globe

3D modeling is useful in many industries and is growing in popularity as a way to maximize efficiency and cost containment. 3D modeling is used in movie production, corporations, architecture and construction, the automotive industry, and many other fields. In the automotive industry, manufacturers are 3D printing auto parts and using 3D printed models to help decide on what improvements need to be made on vehicles.

NASA is currently exploring the use of 3D printing in a variety of ways. Using 3D-printed parts to build rockets and 3D printing food for astronauts are both being researched by NASA. A 3D printer has been sent to the International Space Station aboard an unmanned SpaceX cargo capsule so that, as parts and tools are needed, astronauts can simply print more. NASA has partnered with researchers at Washington State University to investigate the possibilities of 3D printing objects using moon rock.

The medical field is using 3D printing in a variety of ways. 3D-printed organs, made from a person's own cells, can be implanted shortly after being printed. 3D-printed prosthetics have been getting a lot of attention lately, as even high school students are designing and 3D printing prosthetics for both people and animals that need them. Human arms, hands, feet, and skulls are now routinely being printed. Researchers at the University of Toronto have discovered a way of 3D printing human skin tissue, a major advancement in skin grafting. Cornell University is 3D printing ears for children born with birth defects using collagen from rat tails and cartilage from cow's ears. The entire process of 3D printing the ears, including designing the mold and printing the ear, takes about two days.

CSIRO, a technology company, is 3D printing tracker tags to track movements of individual marine species and increase understanding of their behavior. Those species are at risk of extinction from overfishing, so any new technology that can help with gaining that knowledge in a more efficient manner is great. The company has found that the major advantage of 3D printing the tags is that designs can be adapted and tested within a short time period.

Museums are finding 3D printing to be a better avenue for creating models, including those of fossils, for display. Because museums don't display actual fragile fossils, models made from clay have been the standard for years. This process was expensive and extremely time consuming. Now, museums are utilizing 3D printing to create replicas of these fragile pieces for display. Scientists use a 3D scanner to scan the original pieces and create a model from these pieces.

Researchers at Drexel University are taking that a step further by 3D printing dinosaur fossils for museums and using the models to build robotic versions of the dinosaurs to study how they moved and responded to environmental stresses.

3D printing is being used around the world to further industry and science in ways that could not have been imagined before. The powerful impact that 3D modeling and printing is having on the world makes this a fascinating time to be teaching and learning.

Chapter 1 Key Points

The 3D printing process can be divided into two stages: modeling and printing.

The modeling stage is where the user creates a digital model of the object, using software compatible with their particular printer.

3D printing software slices the object into thin layers, and this enables the printer to lay down the material (usually plastic in consumer 3D printers) in layers to build the final object. Melted plastic is extruded into these thin layers according to the *X*, *Y*, and *Z* coordinates set by the design process. The layers begin at the bottom of the object and build the object from bottom to top, layer by layer, until the object is complete.

The most basic specifications that should be considered when deciding on which 3D printer to purchase are print area and volume, print speed, and print surface.

The most commonly used filament in consumer 3D printers is made of plastic. Two types of plastic filament are used in 3D printing: ABS (acrylonitrile butadiene styrene) and PLA (polylactic acid) filament.

ABS filament is petroleum based and comprised of acrylonitrile, butadiene, and styrene, which some research has found to be toxic, especially when heated.

PLA filament is made from plant-based materials. PLA can be made from sugar cane, beet cane, corn, or other vegetable waste and is biodegradable.

There are three stages in 3D printing with students.

1. Students find and print 3D models created by others.
2. Students find, modify, and print 3D models created by others.
3. Students use 3D modeling programs to design and print their own models.

Some of the modeling programs that are fairly easy for students to learn include Tinkercad, 123D Design, 3D Tin, and SketchUp. These programs are covered in detail in chapters 4–7.

There are also student-friendly apps for 3D modeling on tablets. These apps, including Blokify and 123D Design, are covered in detail in Chapter 8.

2

The Role of 3D Printing in Education

Why Integrate 3D Printing into the Curriculum?

3D modeling and printing are both effective tools to build 21st-century skills in students. There are few processes like these that stimulate the imaginations of students and introduce them to the creation and invention of original objects from beginning to end. 3D modeling and printing do more than just provide students an outlet for creativity. Emerging technologies in education such as 3D modeling and printing can easily be utilized to support professional standards and build skills essential in order for students to learn how to become effective users of information.

School librarians have access to powerful guidelines for teaching students through the AASL Standards for the 21st-Century Learner. These standards should be used by school librarians to guide and provide direction for their school library program. The AASL standards include skills, responsibilities, beliefs, and self-assessment strategies that teach students to:

> inquire, think critically, and gain knowledge;
>
> draw conclusions, make informed decisions, apply knowledge to new situations, and create new knowledge;
>
> share knowledge and participate ethically and productively;
>
> pursue personal and aesthetic growth. (AASL 2007)

3D modeling and printing can be used to address all of the above standards and also to provide students practice in critical thinking and problem solving skills. 3D modeling and printing naturally build collaboration skills as students begin to realize that working together during the design and print process is highly beneficial. One of the best things about 3D modeling and printing is the excitement it creates in students.

The International Society for Technology in Education (ISTE 2007) has established standards for technology use for students that can be found on their website at iste.net. ISTE provides standards for effective learning in the areas of (1) creativity and innovation, (2) communication and collaboration, (3) research and information fluency, (4) critical thinking, problem solving, and decision making, (5) digital citizenship, and (6) technology operations and concepts. More information on these standards, as well as the standards for teachers, can be found at iste.org/standards.

3D Modeling and Printing Builds Critical Thinking Skills

According to Paul and Elder (2013), experts in the field of critical thinking, "in a world of accelerating change, intensifying complexity and increasing interdependence, critical thinking is now a requirement for economic and social survival." It is imperative that students build critical thinking skills whenever possible, yet these skills are not practiced often enough in most school settings. Norris and Ennis (1989) provided this simple definition of critical thinking: "Reasonable and reflective thinking that is focused upon deciding what to believe or do." Critical thinking is all about reasoning, reflecting, introspection, purpose, solutions, and the willingness to be wrong. These are all characteristics we hope to encourage and build in all of our students.

The ability to think rationally, process information, and then make decisions based on this processing is difficult for most students. 3D modeling and printing gives students experience in thinking critically about the modeling of an object and how that model will look when the finished product is printed. The modeling process allows students to practice thinking critically and making adjustments as needed based on their results.

According to Robert H. Ennis (1989), successful critical thinkers have several specific attributes. Each of the attributes described by Ennis can be honed in students with 3D modeling and printing. The attributes described by Ennis include the following:

- Capability of taking and changing a position based on evidence
- Being open minded
- Searches for reasons
- Seeks a clear statement of the problem
- Investigates all options
- Uses credible sources

One of the best ways an educator can help students build critical thinking skills is to encourage inquiry. Librarians can cultivate critical thinking skills by encouraging students to question the thinking of others, as well as to question their own thinking. If an educator is easily annoyed by students when they ask many questions, students will stop questioning. Students should learn that asking questions is a great way to clarify or to gain more information. Librarians can teach students how to challenge information in a respectful manner and how to evaluate the credibility of information sources. Being able to critically evaluate information is a valuable skill students will benefit from throughout their lives.

School librarians can incorporate specific activities into the library to help foster critical thinking skills. Providing an online discussion forum concentrating on 3D modeling and printing will allow students to discuss the process of modeling and printing outside the library and give students a means for focused inquiry. Permit students to socialize in the library and have time to discuss 3D modeling and printing, even when there is no 3D modeling and printing going on. Another easy way to promote critical thinking is to encourage students to come up with 3D modeling challenges for each other and give them opportunities to discuss the results. A student, for example, can challenge other students to design a 3D model of a national landmark or a replica of their favorite toy when they were younger. This will also help students view the thinking of others through a different lens while allowing them to take some initiative in designing models.

3D Modeling and Printing Builds Problem Solving Skills

Problem solving is a multilayered process. Clearly defining the problem, brainstorming options, evaluating and choosing an option, and then moving forward with a solution are each important concepts students need to master. 3D modeling and printing both offer opportunities for students to practice these skills in a way that is both interesting and motivating to students.

A great way to embed critical thinking and problem solving skills into the 3D modeling and printing process is to incorporate problem-based learning into the project. Problem-based learning allows students to collaboratively investigate and solve real-world problems. Learning in problem-based projects is self-directed, which allows students to seek their own responses to a problem. An easy way to integrate basic problem-based learning is to look around the school, community, or world and determine a problem that needs to be solved, then have students model and print an object that would help solve that problem.

Often, problem solving is confused with critical thinking and higher-order thinking. Although similar terminology is often used in describing components of critical thinking and problem solving, the concepts are unique. According to Hedges (1991), problem solving is more of a linear process of evaluation, while critical thinking is an overlying set of abilities that allow students to properly facilitate each stage of the linear problem-solving process. Problem solving skills involve recognizing and defining a problem, the ability to comprehend and use concepts and generalizations, testing and revising hypotheses, and finally, forming a conclusion.3D modeling gives students much practice in problem solving. Designing an object, recognizing a problem in the design, and determining ways to fix the design problem are processes that students will encounter again and again in the design process. If an object fails to print correctly, students will use problem solving skills during the redesign process to fix the problem.

3D Modeling and Printing Stimulates Creativity in Students

3D modeling and printing provides students a way to experience creativity in a way rarely seen in schools. Students who think of, design, and then print an object have ultimately used all of their creative juices to go from an idea to a final product they can hold in their hand. This is a powerful process for students and can be the spark that some students need to feel empowered in their own learning.

Creating is the top level of cognition in Churches' Revised Bloom's Taxonomy (2009) and is a compilation of all of the other cognitive levels. Students are required to remember, understand, apply, analyze, and evaluate as they are involved in the creating process. Some great questions to ask students to facilitate the creating process during 3D modeling and printing are:

Can you create a ________ that solves that problem?

Can you create a solution to ____________?

What do you think would happen if ________________?

What is another way to ________________?

In her article *Assessing Creativity* (2013), Susan Brookhart clearly lays out ways that creativity can be evaluated by educators, as well as the criteria for creativity in students. According to Brookhart, creative students:

- Recognize the importance of a deep knowledge base and continually work to learn new things.
- Are open to new ideas and actively seek them out.
- Find source material in a wide variety of media, people, and events.
- Organize and reorganize ideas into different categories or combinations and then evaluate whether the results are interesting, new, or helpful.
- Use trial and error when they are unsure how to proceed, viewing failure as an opportunity to learn.

3D modeling and printing allow students to demonstrate each of these criteria. Because 3D modeling and printing are new tools for most students, it is a whole new learning experience for them. Students are forced to learn, simply by participating in the modeling and printing process. Because each new object modeled by students is different, although they may have gone through the process before, it will be a different and new experience every time for the student.

Students who are engaged in 3D printing begin to actively seek out creative opportunities to utilize the printer. They become excited about the technology and, once they see the possibilities, they begin to bring in new ideas for ways to use the printer. Often, students will get an idea from what they are learning in the classroom, and that becomes the seed for their next 3D printed object. They begin to pay more attention to news reports or online articles about 3D printing and are excited that they are a part of that same process. Students will often find examples from the Internet of how 3D printing is being used to better the world around them and share this information with other students. Creating a spot in the library where students can post these articles will encourage the excitement to continue. Providing students with ways to share information will spur creativity. This will also build a shared synergy between students, which will only improve the creative process.

One of the largest barriers to creativity in students during the 3D modeling and printing process can be the student's view of failure. Unfortunately, years of traditional education have trained many of our students to be fearful of failure. In their eyes, failure is not a good thing; it only leads to bad grades. This is challenging to overcome, and school librarians should know that overcoming the fear of failure will probably be a slow process with most students. Show students that with 3D modeling, failure is seen as a positive thing because it means students are able to evaluate and reengineer their designs. In the library, second chances (and third, fourth, fifth chances) are great opportunities for learning. In the makerspace in our school library, we celebrate the second, third, and fourth attempts even more than the first tries. Failure *is* an option and, in fact, it is a preferred option. Students learn that failing is OK when it is in a safe place where no one is judging them. When students stick to it, remodel, and reengineer, they are busting through their preconceived notions of what failure means. School librarians can help students overcome the fear of failure by stressing the process over the outcome. Students will begin to recognize the importance of the process while still being excited about the object they can hold in their hands in the end.

3D Modeling and Printing Naturally Encourages Collaboration

Collaboration is no longer merely a suggestion for the learning process, it is absolutely essential, for both educators and students. Nothing encourages collaboration between teachers like being thrown into something new, and students are the same way. Technology that students are unfamiliar with provides new opportunities for students to collaborate. 3D modeling and printing is no exception. As with anything, you will have some students who understand the concept of 3D modeling before other students do. Regardless of the age of students you work with, students who grasp the concept first will help you teach it to other students. The great thing about this is that this is usually not anything the school librarian has to teach. It will most likely happen naturally on its own. If the school librarian does end up needing to promote collaboration between students, it is fairly easy to do by having older or more advanced students work with those who need more assistance. Students who struggle with 3D design will develop their skills and will soon be able to assist and teach other students.

Collaboration among students at the same school is great; but if the school librarian can promote collaboration not just among students, but across schools, the district, and the community, even better. Perhaps different grade levels on a campus or on different campuses can collaborate to design and print 3D models. Fifth-graders can assist third-graders, even if they are on another campus, through FaceTime or Skype. Students from an engineering or modeling class at a local high school can teach students at a nearby elementary school how to use modeling software.

Community members or parents can volunteer to help students learn about the engineering process. Investigate businesses in the community to find those that utilize 3D printers as part of their work and ask them to visit or Skype in to talk about how they use some of the same skills that students are learning. Students will love seeing that the skills they are learning now will translate into career opportunities later on. Connecting the 3D modeling and printing process that students are involved in to what is going on in the real world will build a bridge to the future for many students. Think beyond the walls of your library to establish collaboration opportunities.

3D Modeling and Printing Sparks Excitement of Learning

One of the major benefits of integrating 3D modeling and printing into the curriculum is that students get excited about school and learning. Both students who have been successful in school and those who haven't will be eager to work with this technology over and over again. 3D modeling and printing may be just the thing to rekindle the magic for students who have become disenchanted with school. Many schools have found that after adding a makerspace with a 3D printer to the school library, library traffic increased, which led to more books being checked out. Many schools report that after building a makerspace, student attendance rates increased. Students who never considered school attendance a priority all of a sudden find a reason to come to school. As one student said after telling the school librarian his mother didn't want to get up to take him to school, "I told my mom I needed to come to school because I wanted to print my robot on the 3D printer today." Seeing students experience the excitement and joy of learning is why many of us were drawn to education in the first place, and 3D printing gives students these opportunities to shine. 3D modeling and printing engage students as some of them have never been engaged before, but one of the best things about 3D modeling and printing is how they empower students to gain a deeper understanding of new content.

In his article "Turning on the Lights," Marc Prensky (2008) discusses the importance of reaching students by bringing opportunities typically given to students in an after school setting into the classroom on an everyday basis. School libraries are the perfect place for this to happen by surveying students to see what types of projects and challenges they would like to experience, giving students opportunities to connect with others across the globe, and understanding the goals students have and helping them meet these goals. 3D modeling and printing will certainly turn on the lights for many students who have become disenchanted with school and learning.

3D Modeling and Printing in School Libraries Provides Students Access to Emerging Technology

There may be a few elementary and secondary students who have access to 3D printing technology at home, but most do not. Students may have heard of 3D printing but never seen a 3D printer in real life. Some students may not have ever heard of 3D printing. A benefit of purchasing a 3D printer for the school library is giving students access to technology they wouldn't have otherwise. Exposing students to cutting-edge technology will benefit them the rest of their lives. As a school librarian in a Title 1 school with around a 90 percent low-income student population, I have always felt a responsibility to not only expose students to this type of technology, but to help them learn skills to use the technology to get a leg up on job competitors after high school and college. I strongly believe this responsibility should be taken seriously by educators, regardless of student demographics.

3D Modeling and Printing Increases Spatial Intelligence in Students

Spatial intelligence is what allows us to visualize and manipulate 3D images and shapes mentally. Spatial intelligence is also used when putting together a puzzle or looking at a map. Students with good spatial intelligence are good visual thinkers. These are also the students that don't find Rubik's Cube very challenging.

Spatial reasoning is the skill to generate, rotate, change, and manipulate images mentally, and many students have not had much practice in developing this skill, even though it is an extremely important skill for students to have. Jonathan Wai, David Lubinski, and Camilla P. Benbow (2009) studied the importance of spatial reasoning in the future learning and work of adolescents. The study found that good spatial reasoning skills were an important factor in the adolescents going on to work in a STEM (science, technology, engineering, and math) field in the future and to pursue advanced academics in a STEM field. Because spatial reasoning is not usually used in schools to test for gifted and talented programs, these researchers suggest that students who excel in spatial reasoning but not in other areas are often overlooked for STEM programs. This is where school librarians can help.

3D modeling and printing is a great avenue to allow students to practice spatial reasoning. Designing using 3D modeling software requires students to generate shapes and then manipulate them so that the result is the intended design or object. Students learn to visualize what the final object will look like and will develop the skill to change designs to make the object look like what the student envisions. Spatial reasoning is difficult for some students, so allowing them plenty of time to manipulate objects as many times as needed is imperative.

3D Modeling and Printing Allow Students to Experience Success

Often, students drawn to opportunities to be creative and to technologies such as 3D modeling and printing are the same students who struggle in the classroom. Whether their challenges are academic or behavioral, these students have challenges in the traditional classroom setting. 3D modeling and printing gives these students the opportunity to excel in an area of learning that allows them to experience success, because success is measured differently in 3D modeling and printing. Being able to model, redesigning as necessary, and successfully 3D printing an object of their own creation can give students a feeling of success, accomplishment, and satisfaction in a way they may not have experienced before.

3D Modeling and Printing Can Teach Multiple Literacies

The importance of multiple literacies has been a hot topic in education recently. Literacy used to be seen simply as reading but now is understood to include a broad range of skills. The 2007 publication *Standards for 21st Century Learners in Action* by The American Association of School Librarians (AASL) lists four types of literacies: visual, digital, textual, and technological. Now more than ever, students need to be adept at finding and evaluating information, using information ethically, and connecting globally with others in a safe manner. 3D modeling and printing is an effective way for students to experience multiple literacies, especially visual, digital, and technological literacies.

3D Modeling and Printing Provides a Way to Integrate Technology

Integrating technology into the curriculum promotes active engagement and empowers students to learn skills for the 21st century. There are several models for technology integration, one of which is the SAMR (Substitution Augmentation Modification Redefinition) model. The SAMR model is a guide for educators utilizing technology as part of the teaching process and describes four levels of technology integration. The goal is that educators focus on the top two levels of technology integration, Modification and Redefinition. These levels are the ones in which technology is not simply substituted for what used to be done, but is used to provide students an opportunity to do new tasks that were not even possible before. 3D modeling and printing provide students with these opportunities to create and invent like never possible before.

For more information on the SAMR Model, visit hippasus.com/rrpweblog/

3D Modeling and Printing Provides an Opportunity for Assessment

Ideally, in a library makerspace area, learning and assessment would be student driven. Students would have a choice in deciding what they want to build or design and would be able to revel simply in the joy and wisdom gained from designing and printing the project. In reality, to be able to rationalize the inclusion of a 3D printer into the library program, the librarian may need to have a plan for assessing student learning. Self-assessment is a good way to incorporate assessment into a 3D modeling and printing program. One of the best assessments is to have students write about the modeling and printing experience. What worked and what didn't? What did students learn from this project that can help them in the next one? A Google form works well for student self-assessment and can give students the opportunity to think critically about their experiences. Students can also document their experiences in a video or multimedia presentation that can be shared with other students, parents, and school staff.

Chapter 2 Key Points

Librarians should be familiar with professional standards such as the AASL Standards for the 21st-Century Learner and the ISTE Standards for Students.

Incorporate activities that specifically focus on building critical thinking skills in students, such as encouraging students to come up with 3D modeling challenges for each other and giving them opportunities to discuss the results.

In 3D modeling and printing, students should know that failure *is* an option and, in fact, is the preferred option.

Promote collaboration by connecting students across the school district. Build relationships with community members, parents, and local businesses that may already utilize 3D printing or are willing to learn.

3D modeling and printing may be just the thing to rekindle the magic for students who have become disenchanted with school. Ensure these students are a part of the program.

3

3D Modeling Programs and Applications

3D modeling programs and applications allow users to create a design blueprint of the object to be printed. Although there are similarities between modeling programs, each program has its own interface and way the design process is navigated. While programs have some similarities, they also have many differences. Programs have individual application strengths, as well. For example, modeling software programs designed for gaming usually aren't especially good for basic 3D design and modeling.

A variety of modeling software programs and applications can be used by students for 3D design and modeling, and these programs and applications vary greatly in their complexity and ease of use. The majority of the programs and applications listed have tutorials on YouTube, which make learning these programs quite easy, even for beginners. Although there are some modeling software programs available for a fee, most of the programs mentioned in this chapter are free. Some of the programs and applications that are free to download contain paid content inside the program. Sometimes the content for basic use is free, but users wanting more than the basics would need to look into purchasing access to more-complex design tools. There are a few applications at the end of this chapter that are not free but are still solid choices for 3D design for students.

One of the primary aspects to consider when choosing a 3D modeling software program for student use is how easy it is to learn and how intuitive it is for users. Especially when students are new to the 3D modeling process, it is important to choose software that students can learn somewhat easily and enjoy using. This will help reduce frustration levels of students while learning to design in 3D software. Once students get the hang of 3D modeling, they can be introduced to more-sophisticated software programs and applications. Each program mentioned in this chapter includes a brief description of the program and information about its use. This list is not meant to include all of the modeling programs available but will hopefully provide librarians and students who are new to 3D modeling a place to begin.

There is a variety of 3D modeling programs and apps that students can use to learn modeling basics. Depending on the type of equipment available (laptops, iPads, Chromebooks, etc.), librarians can choose the modeling programs that are just right for their students.

3D Modeling Programs

3D Crafter

amabilis.com

3D Crafter is a free program that older students or those looking for a challenge may enjoy. Although not quite as intuitive as other 3D modeling programs, 3D Crafter includes tutorials for all levels of users. 3D Crafter has animation capabilities, allowing users to create animated scenes by positioning designed objects within a frame and then viewing them in real-time. The scenes can also be recorded and saved.

3D Tin™

www.3DTin.com

3D Tin and the 3D Tin logo are registered trademarks or trademarks of Autodesk, Inc., and/or its subsidiaries and/or affiliates in the United States and other countries.

3D Tin is a free, browser-based modeling software program. While 3D Tin does have limitations that may frustrate more-advanced users, it offers good basic design features for students new to 3D modeling. Users click on the shapes in the toolbar and then add them to the plane to create an object. While not quite as easy as Tinkercad's drag-and-drop design process, students of all ages should find this modeling program fairly easy to learn. Learn how to use 3DTin in more detail in Chapter 6.

Anim8or

anim8or.com

Anim8or is a 3D modeling and animation software program that is free to download. In addition to 3D modeling, this program also allows users to build 3D scenes and convert them to .AVI movie files or .JPG or .BMP images. There are some tutorials on the Anim8or website, and there are additional tutorials that can be accessed on YouTube.

Blender

Blender.org

Blender is a free 3D modeling and animation program that is good for students who have some experience with 3D design as it may be a little overwhelming for younger students or those new to 3D design. The Blender website contains a vast number of tutorials to help users learn the program. Blender has a variety of capabilities suitable for more-experienced 3D modelers, such as 3D game creation. Creating 3D games in Blender is a great way for students to experience 3D modeling in a creative and exciting way.

freeCAD

freecadweb.org

freeCAD is a good, basic program that offers 3D design along with capabilities for motion simulation. This program is great for students who are just learning 3D design as well as those interested in learning more about motion simulation. freeCAD allows users to create and manipulate 3D solid parts that can be connected using joints, springs, motors, and more.

Inkscape™

inkscape.org

Inkscape is an open source program, which means the original source code is available and can be modified by anyone. This program is similar to Adobe Illustrator. Although Inkscape doesn't have the full range of features that Illustrator has, it is still a powerful vector graphics illustrator that can be used for 3D modeling.

K-3D

K-3D.org

K-3D is a free 3D modeling and animation program that can be downloaded from the Download section of the website. Users can model, animate, and interact with animations while they play the videos for maximum productivity. K-3D is great for artists utilizing 3D modeling software because of the flexibility in editing. Users can easily edit, revise, and return to original models by using K-3D's undo tree, which gives users the opportunity to easily retrieve older revisions of a model, which is incredibly helpful in the design process.

K-3D is great for polygonal modeling, and also includes tools for patches, curves, and animation.

LEGO Digital Designer

ldd.lego.com

Regardless of the age of students, LEGO Digital Designer is a popular choice for 3D modeling. This free program allows users to create objects with virtual LEGO bricks and is very simple to learn and use. The design grid includes three areas: tools on top, bricks and pieces on the left side, and the center grid where the design takes place. This program will be a favorite of students of all ages.

Now Make This

Nowmakethis.com

This site touts itself as a "creative block-building social game" and is definitely more of a game than a modeling program. This website provides younger students with a fun way to learn the basic idea of 3D modeling. Students simply click on the flashing PLAY button at the bottom of the page and they can make objects or guess the names of objects created by others. Users are given three choices of new objects to create and are shown six tiles they can use to make that object. Once finished, students are able to enter their first name and share the model with others.

OpenSCAD

Openscad.org

OpenSCAD is a free software program for creating solid 3D models that is available for Linux/UNIX, Windows, and Mac OS X. This program works differently than others because the user creates a script that describes the model, and OpenSCAD reads the script and builds the model. OpenSCAD focuses on the script programmed by the user and would probably be best utilized by secondary students who are ready for a challenge in 3D modeling.

> The Makerbot site contains some fantastic lessons on OpenSCAD at http://www.makerbot.com/tutorials/openscad-tutorials/

Printcraft

printcraft.org

Printcraft combines two of students' favorite things: 3D printing and Minecraft. The benefit of this program is that it utilizes the love students have of the Minecraft video game and the existing skills students have acquired through learning and playing the game. Printcraft allows students to turn their Minecraft models into 3D printed models they can hold in their hand. Models can be created and printed in three easy steps: (1) log in to a Minecraft server, (2) build an object, and (3) press the Print button to open the Printcraft model page where the model can be printed.

Sculptris

pixologic.com/sculptris

Sculptris is a fun and easy 3D modeling program that students will enjoy learning to use. Sculptris allows users to create objects by virtually sculpting a sphere of clay. Sculptris doesn't have as many features as Pixologic's pricier ZBrush program, but it is more than suitable for student use.

Seamless 3D

Seamless3D.com

Seamless 3D is a 3D modeling and animation software program that can be downloaded for free. Users utilize "build nodes" to create objects, and the site includes tutorials and a link to a discussion forum for users.

SketchUp

Sketchup.com

SketchUp has several different products for 3D modeling, and SketchUp Make is the free version of their program. SketchUp Pro is free for educators, and the company provides a discount for student and school use. SketchUp's website includes the 3D Warehouse, which contains many models that have been created using their program. SketchUp Make is a simplified version of the program that is good for younger students or beginners. After students get more comfortable with the Make version, they could move to the more advanced Pro version. A benefit of SketchUp is the plentiful resources, training, and video tutorials available on the website. Learn more about SketchUp in Chapter 7.

Sweet Home 3D

Sweethome3d.com

Sweet Home 3D is a different type of 3D modeling program that students find engaging and fun. This program is a free interior design program that gives users the choice of downloading the program to a computer or using it online. Users can create a home or building by adding walls, doors, windows, and furniture.

Although designed in 2D, users can view their creation in 3D, either in aerial view or by navigating through the structure as if they are walking around in it. Challenge students to use Sweet Home to design famous homes in literature, such as the home of the three bears or the witches' house in Hansel and Gretel.

Tinkercad™

Tinkercad.com and Free app on iTunes

Tinkercad is a free, easy-to-use 3D modeling program that is browser based, which means it can be accessed anywhere there is Internet. This program uses a variety of shapes as the basic building blocks for designing objects to be 3D printed. Students can add or remove material from shapes and are able to upload their shapes. The program provides guided lessons that are easy to follow, even for younger students. The lessons take students through the three basic steps of modeling in Tinkercad, which are (1) placing shapes, (2) adjusting the size and colors of shapes already placed, and (3) combining shapes to create more-intricate designs. To use Tinkercad, students will need to create accounts, but this is a very simple process using only a name and e-mail address. Students can log in to Tinkercad from any location to access their accounts and designs. Designs are saved, so students can pick up designing where they left off. Learn more about Tinkercad in Chapter 4 of this book.

The Tinkercad, 123D Design, 3D Tin, and SketchUp modeling programs are covered in detail in chapters 4 through 7. These chapters provide basic design how-to's for beginning modelers. As students engage in modeling and become more familiar with the process, they will naturally want to move beyond the design basics. Students will experiment and try new things while designing, and this should be encouraged. Students will learn and will begin to teach each other. The detailed chapters will provide support for librarians new to 3D modeling and printing. After designing an object or two, modeling will become easier. It's definitely a process that gets better with practice.

Applications

> To disable in-app purchases on the iPad, enable Restrictions and make sure In-App Purchases is turned off.

Applications, or apps, are an easy way for students to engage in 3D modeling using tablets. Many schools are adding iPads or iPad Minis to their inventory because they are a relatively inexpensive way to add mobile technology to the library and classroom. Most of the 3D modeling apps described are free but offer in-app purchases.

For more detailed information on several 3D modeling apps appropriate for student use, see Chapter 8.

Free 3D Modeling Apps

Autodesk 123D collection of apps

http://www.123dapp.com

Autodesk maintains a comprehensive website that offers detailed directions for use with all of the apps and programs below.

123D Catch™

Free on iTunes and Google Play

This app allows users to 3D print an object using a series of photographs. Users start a new "capture" by walking around an object while taking overlapping photos at varying heights. For successful results, users are encouraged to take 20–40 photos, ensuring there is even lighting around the object and that there is contrast between the object and the surface the object is on. After the object is prepared for printing, users can print the object themselves or have it printed through a 3D printing service for a fee and shipped to them. 123D Catch can be utilized as an app on an iPad or through a program downloaded to a PC.

123D Design™

Free on iTunes

123D Design is a 3D modeling and editing program that can be downloaded and used on a laptop or iPad. This app provides good basic design tools with built-in tutorials and quick tips to help make designing easier. The completed design can be exported and then 3D printed.

123D Sculpt™

Free on iTunes and Google Play

Users can sculpt and decorate virtual clay to make a 3D model that can then be 3D printed. As with the other Autodesk apps, tutorials are built in to make sculpting easy for students of any age. Users can select creatures, geometric shapes, or objects as their virtual clay shape to sculpt. To sculpt or paint in the app, users drag a finger across the shape. Dragging two fingers on the empty space around a sculpture allows the user to move the sculpture around.

123D CreatureShow™

Free on iTunes

CreatureShow allows users to manipulate provided creatures by moving dots, or sensors, to change the creature as the user wishes. Users can then change lighting, add special effects, and then save the creature or share it with others.

123D Creature, CreatureShow, Catch, Design, and Sculpt are registered trademarks or trademarks of Autodesk, Inc., and/or its subsidiaries and/or affiliates in the United States and other countries.

Blokify

Free on iTunes with in-app purchases

Blokify is a 3D modeling app easy enough for the youngest of users. Design is based on bloks and has a Minecraft feel, which is enticing to fans of the game. Students drag and drop the bloks to create an object, which can then be sent to a 3D printer. This app makes 3D modeling fun by adding a variety of backgrounds and music. Blokify has many in-app purchases, so educators will want to ensure that in-app purchases are restricted before students begin to create their own designs. After a design is created, it is e-mailed to you in the .STL format, so the object can be 3D printed immediately.

Makers Empire

Free on iTunes and Google Play

Makers Empire is a new app for 3D modeling on tablets. Users can design a character, draw on and customize 3D objects, create new 3D objects and add text, and build objects with blocks, bricks, and cubes. Objects can be exported as .STL files, but this requires an in-app purchase.

SpaceDraw

Free on Google Play for Android only

User manual available at http://www.scalisoft.net/Spacedraw_manual.pdf

This app provides Android users with an impressive 3D modeling program. Students beginning to learn 3D modeling may find this app frustrating at first but, with some time and effort, can learn to use it to create great 3D objects.

Tinkerplay

Free on iTunes

Tinkerplay (formerly Modio) allows users to select body parts and templates and drag and drop the pieces to create any character, human or animal. Heads, torsos, and limbs are interchangeable pieces that click into place and can be added or removed by simply dragging the piece. Body parts can be enhanced with textures and shapes. Finished designs can be 3D printed, and although Tinkerplay is optimized for Makerbot brand 3D printers, it also works well with other brands of printers. With the press of a button, the model is placed on a virtual build plate.

Paid 3D Modeling Apps

Artist3D–Modeling Tool

$9.99 on iTunes

Artist3D is easy to learn and use. Objects can be moved and scaled easily simply with a swipe of the finger. Objects can be sketched by drawing an outline and can be painted. Artist3D works with the Shapeways 3D printing service.

Let's Create! Pottery

$4.99 on iTunes and Google Play

This app allows students to experience the concept of 3D modeling in an artistic way. This app has a good beginning tutorial. To create a pottery piece, users move their finger on the piece of clay. Students are able to move their finger up and down and from side to side to make the pottery piece taller or shorter, and wider or thinner. After the piece is fired, the user decorates it. This app is a great way to combine 3D modeling and printing with fine arts.

Verto Studio 3D

$13.99 on iTunes

Verto Studio 3D is a fully functional 3D modeling app that has as many capabilities as most desktop and browser-based programs. Verto Studio 3D supports importing, exporting, viewing, and editing 3D models of any size.

Chapter 3 Key Points

A variety of modeling software programs and applications can be used by students for 3D modeling, and these programs and applications vary greatly in their complexity and ease of use.

The majority of the following programs and applications have tutorials on YouTube, which make learning the program quite easy, even for beginners.

Especially when students are new to the 3D modeling process, it is important to choose software that students can learn somewhat easily. Once students get the hang of 3D modeling, they can be introduced to more sophisticated software programs.

There is a variety of 3D modeling programs and apps that students can use to learn modeling basics. Depending on the type of equipment available (laptops, iPads, Chromebooks, etc.), librarians can choose the modeling programs that are just right for their students.

4

Focus on 3D Modeling: Tinkercad

www.Tinkercad.com

Probably the easiest 3D modeling program for students to use, Tinkercad is a free, web-based program that is simple to learn for even young students. Tinkercad is a good beginning program for students who have absolutely no experience in digital modeling.

3D modeling is similar to many other skills that are best learned by doing. Don't be afraid to open up these programs and get to know them. You can't break them, and the more you do with it, the easier it will become and the more sense it will make. One of the great things about Tinkercad is that while it is easy to learn, the skills learned by using Tinkercad can be transferred to most other 3D modeling software programs.

Tinkercad can be operated on computers running Microsoft Windows Vista or newer, Apple OS X 10.6 or newer, or Google Chrome OS on a Chromebook. Access is easy as Tinkercad is web-based and does not need to be downloaded. It is best to use the Google Chrome 10 or newer or Mozilla Firefox 4 or newer browsers to access the Internet if you are going to be using Tinkercad. Both Chrome and Firefox support WebGL, or Web Graphics Library, which simply put means that Chrome and Firefox contain a library of images that allows users to create and manipulate 3D objects without using plugins.

If accessing Tinkercad on a Mac using Safari, WebGL has to be enabled before accessing the website.

To enable WebGL in Safari:

1) Open the Safari menu and click on Preferences.
2) In the Preferences window, click on the Advanced tab.
3) Check the Show Develop Menu in the Menu Bar box that is located near the bottom of the window.
4) Open the Develop window in the menu bar.
5) Select Enable WebGL.

If you are using Chrome or Firefox and you are getting a message that WebGL is not enabled or the browser is not working with WebGL for some reason, close the browser and try accessing Tinkercad using another browser. For example, if you are using Chrome and are getting the WebGL error message, close Chrome and try using Firefox.

If you are using a browser that does not support WebGL, you will get the following message:

Create free Tinkercad account

Tinkercad requires HTML5/WebGL to work properly. It looks like your browser does not support WebGL.
Troubleshoot WebGL support.

Let's begin learning the basics of Tinkercad! To log in to Tinkercad, go to www.Tinkercad.com. The site will look something like the figure below, although the model shown on the left side of the screen will probably be different as different models created by Tinkercad users are showcased.

Next, click on the blue Sign Up For Free Account button at the top right of the screen.

To sign up for an account, users enter their name, e-mail address, and create a password that is at least six characters in length. Users then enter their date of birth, and then review the terms and privacy policy before clicking the Sign Up button. Students under 13 years old can create an account with parent or guardian permission. When creating an account for students under 13 years of age, users are prompted to enter a username, password, and parents' e-mail address.

For younger students, librarians can create a library account and have students log in to that account, create a model, and save the design in that account. Tinkercad is creating educator accounts that will be available soon. These accounts will help librarians immensely by providing classroom management tools and easier account creation for students under 13 years old.

Users over 13 who use an e-mail address to create an account are able to edit their profile by clicking on their username in the top right corner of the screen. Users are able to add a photo to their profile, but librarians should be cautious about allowing students to do so. For safety and security reasons, students should *not* add photos of themselves to their profiles. If students want to add photos to their profile and it is within school policy, instead have them create an avatar or add a generic image. Users are also able to add information about themselves to their profile but students should not add any personal identifying information. Students should be allowed to experience profile setup because most online programs have this feature and students benefit from learning how to engage in this process in a safe manner. Librarians can use this opportunity to teach digital citizenship skills by prompting students to discuss why adding personal information to profiles is not a good idea for children and teens. Librarians can discuss why photos should not be used and let students have a choice in what they post, as long as it is within the guidelines and policies of their school and school district.

Guided Lessons

After logging in for the first time, users will see four step-by-step lessons that can be used to get to know Tinkercad and how it works.

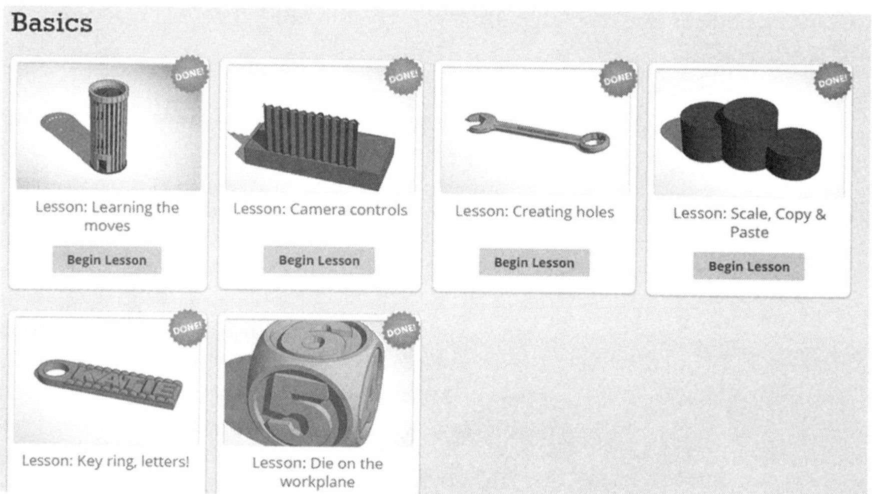

The four lessons are:

- Learning the Moves
- Camera Controls
- Creating Holes
- Scale, Copy and Paste

To work through a lesson, users should click on Begin Lesson for each of the four lessons. To complete more lessons, users can click on the Learn tab at any time.

Tinkercad includes helpful lessons on a variety of topics. These guided lessons are fantastic ways to introduce students to the world of 3D modeling. These step-by-step lessons provide new users with guided practice at designing and creating models using Tinkercad. These lessons are a great way to provide students guided opportunities to design 3D models and learn how to use the tools in Tinkercad.

Specific modeling projects with lessons include:

Basics:

- Key Ring, Letters!
- Die on the Work Plane

Accessories:

- Minecraft Party Glasses
- Cufflinks

Gadgets:

- Luggage Tag
- Chess Pawn
- Saw Shaped Wrench
- Money Clip
- Die From Scratch
- Ruler - cm

Buttons:

- Basic Button
- Duffel Button
- Flower Button
- Heart Button
- Bat Button
- Skull Button
- Teddy Button

Jewelry:

- Simple Heart Ring
- Diamond Ring
- Basic Ring
- Easter Ring
- Cylinder Earrings
- Twist Earrings
- Hexagonal Earrings
- Box Earrings
- Flower Necklace
- Bracelet
- Simple Chain
- Advanced Chain

Miniatures:

- My First Boat
- Model Train Body

Figures:

- Gear of Beach Bunny for Lesson
- Kraken
- Robot

Home Décor:

- Cupboard Knob
- 60 Minutes Clock
- Clock for Laser Cutting

Tinkercad Menus

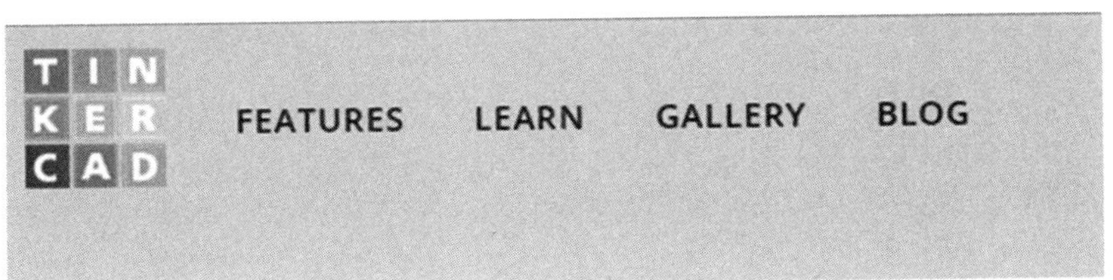

Tinkercad has three menus that allow users to easily access models and designs.

Dashboard menu – Opening this menu allows users to search designs, create new projects and designs, see collections, and view (and follow) Tinkercad's twitter feed. Users can click on All Designs under the word Collections to see all models they have designed themselves and models they have modified or copied from other Tinkercad users. Clicking on Create Project permits users to group 3D models together. This would be helpful if the user is designing several different models as part of one project.

Gallery menu – This menu permits users to view 3D models that have been designed and saved by other users. Models that are most popular can be viewed by clicking the Hot Now tab. This will show models that are being viewed and printed the most. The Newest Things tab shows designs most recently uploaded by users. Tinkercad staff members select their favorite designs and users can access these under the Staff Favorites tab. The #template tab allows users to see 3D design templates on Twitter that have been posted by Tinkercad users.

Learn menu – Clicking on this menu takes users to guided lessons on using and creating Tinkercad models.

Search

At the top of every Tinkercad window, there is a search box that allows users to search the site. Users can search for models by entering descriptive terms for what type of model the user wants to find. For example, users could search for superheroes, animals, and jewelry, or more specifically for Batman, cats, and rings. Users can also use the search box to search the names of Tinkercad users to find all of the designs created by that user.

Edit Profile button – clicking this button allows users to edit the photo or information in the user's profile.

Sharing Content

When users create an account, the Terms of Service state that the user agrees that their designs can be reproduced and modified. Because designs can be seen by anyone searching in Tinkercad, it is important for students to know that there are common-sense rules to follow. Show students the Terms of Service to teach them to actually read the Terms of Service when creating accounts and to show them what is allowed and what is not.

Copying Models Designed by Other Users

Copying models created by other users is the easiest way to get started using Tinkercad. To find a model to copy, the user should use the search box at the top of the screen. Once a model the user wants to copy is found, the user simply clicks on the picture of the model. This will open the model into a larger photo along with buttons.

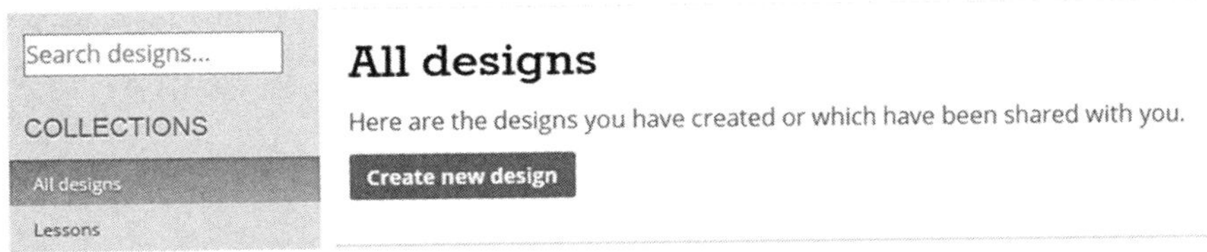

Clicking on the View 3D button opens the design in a 3D view. The user is then able to click and rotate the object in a 360° view to see the object from all sides. Clicking on the picture of the model takes the user back to the non-3D view of the design.

If the user wants to copy and 3D print the model just as it was designed by the original user with no changes, the user can click the Download for 3D Printing button. If the user would like to take the model and make changes, the user can click on blue Copy & Tinker button. This will open that model up on the design grid so the user can make changes to the original design. This will allow the user to use the design tools in Tinkercad, which is covered in the next section.

Designing a New Model in Tinkercad

The autosave feature in Tinkercad automatically saves models during the design process. Users can also save by clicking on the Design tab and then Save, but it is not absolutely necessary, as Tinkercad automatically saves.

To begin a new design in Tinkercad, click on the blue Create New Design button on the Dashboard menu.

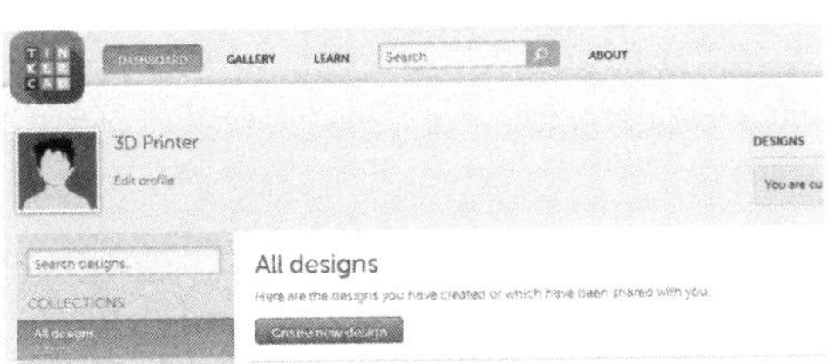

Clicking on the Create New Design menu opens up the workplane. The workplane is the plane on which the model is designed.

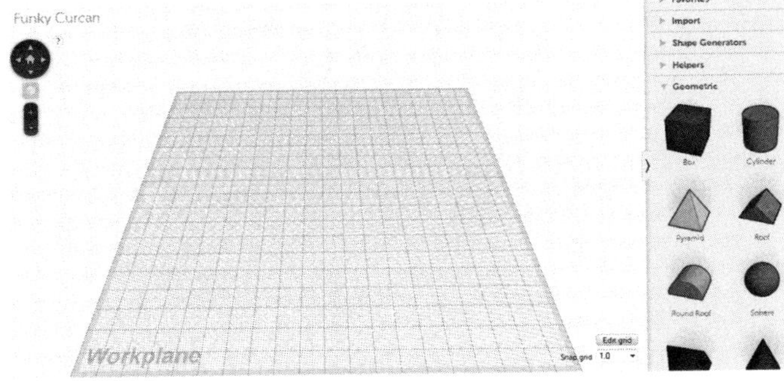

The + and – buttons to the left of the workplane allow the user to zoom in or out. Sometimes it is beneficial to decrease the size of the workplane so the entire workplane and design can be seen. To do this, simply click on the – button until the workplane is the size wanted. To zoom in to see small details, click the + button until the details can easily be seen.

The rotate button allows the user to rotate the workplane in all directions. Rotating the workplane changes the user's view of the design.

Selecting the Edit Grid button allows the user to change the units of measurement on the workplane. Users can choose between millimeters and inches, and switch between these at any time. This button also permits users to change the size of the workplane. For beginners and for users without one of the specific 3D printers listed, the default workplane size will work just fine.

When the workplane is opened, a new set of menus will be displayed at the top.

Tinkercad automatically assigns each new design a name, which has absolutely nothing to do with the actual design itself and sometimes can be quite silly, which students love. The name of the design in the figure above is Funky Curcan. This can easily be changed to a name that better describes the design. To change the name of the design, click on the Design menu at the top of the screen, and then on Properties. The Properties window will allow users to change the name of the design, to change the design visibility from Private to Public, and to change the License type. Clicking on each option prompts brief explanations of each choice to help users decide which to select.

To begin a design, find the geometric shapes on the right side of the workplane. Click on the Geometric tab on the right side and this will open up the Geometric Shapes that can be used to create a design.

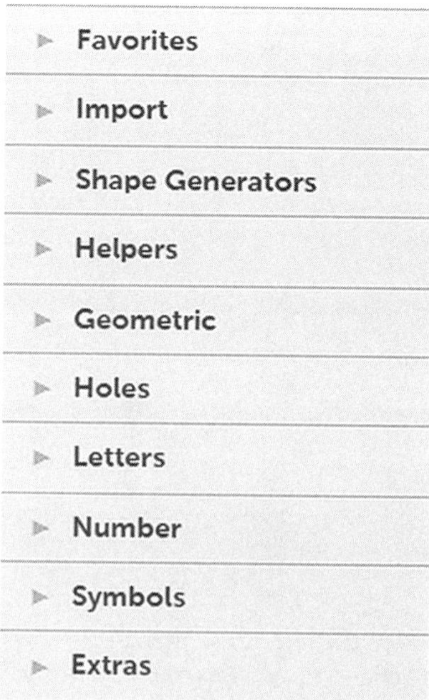

To move a shape to the workplane, click and hold the shape and continue holding it while dragging the shape to the workplane. When the shape is on the workplane, release the click and the shape will stay on the workplane. The color of the shape can be changed by clicking on the shape and then clicking on the color box inside the Inspector box. Changing the color of the shape will not affect the color of the actual model that is 3D printed. If a shape is red on the workplane and the 3D printer is printing using blue filament, the shape will be blue when printed.

To activate controls to change the height, width, and length of the shape, click on the shape. Look closely at the figure below. Controls for the shape include white squares that show measurements, a rotating arrow, and an arrow on top of the shape.

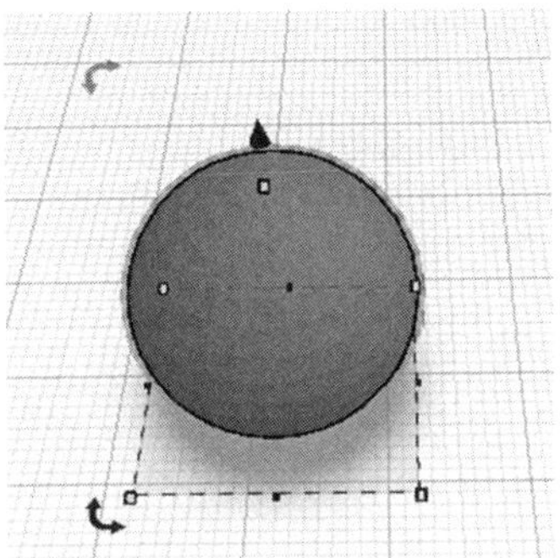

Clicking and holding any of the white squares while dragging will change the measurements of the individual shape. Remember that measurements can be shown in inches or millimeters and can be changed by clicking on the Edit Design button at the right bottom of the workplane.

The Tinkercad workplane includes an Undo button, just like many other programs users may be familiar with. Students are encouraged to take risks while designing just to see what will happen or what something will look like. The Undo button can be a 3D model designer's best friend.

The white square at the top of the shape indicates the height of the shape. Clicking and holding the white square at the top of the shape while dragging up or down will change the height of the shape. Clicking and holding white squares at the corners of the shape while dragging will change the length and the width of the shape.

Sometimes a whole shape is not wanted in a design. Users may not see the shape they need to create the model they have in mind. This is where the Hole features come in handy. The Hole tool is located in the Inspector box that appears when the user clicks on an individual shape.

Tip: This is a good exercise for students to learn how to control the size of objects is to describe certain shapes with specific dimensions they can create. For example, ask students to create a cube with dimensions of 3″ x 3″ x 3″.

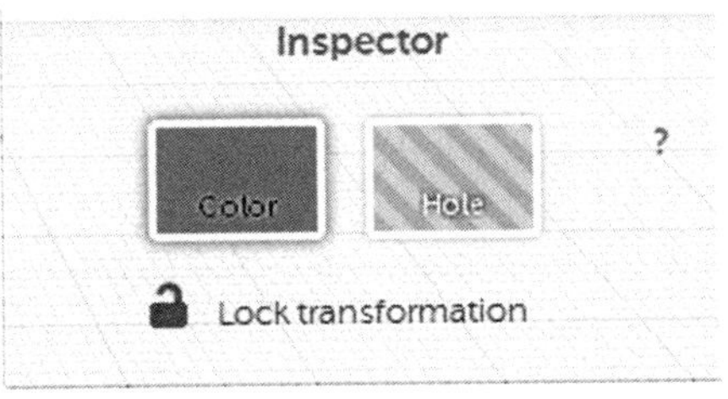

The Hole tool can be used to create a hole, or hollow, inside another shape. This tool allows the user to create unlimited shapes and designs. Clicking on the Holes tab on the right side will allow users to choose between two hole shapes: box and cylinder.

- Favorites
- Import
- Shape Generators
- Helpers
- Geometric
- Holes
- Letters
- Number
- Symbols
- Extras

Holes can be added to the workplane in the same way a shape is added. Click, hold, and drag either a box or cylinder hole shape to the workplane. Hole measurements can be changed in the same way shape measurements are changed. When the hole measurement is the size wanted, drag the hole onto a shape. The Hole tool can be used again and again to refine the shape to make a custom shape. It can be useful to use two different colors for shapes that are being combined to make individual shapes easily distinguishable as they are being combined. Next, use the group tool to combine the hole and shape to make a custom shape.

Grouping

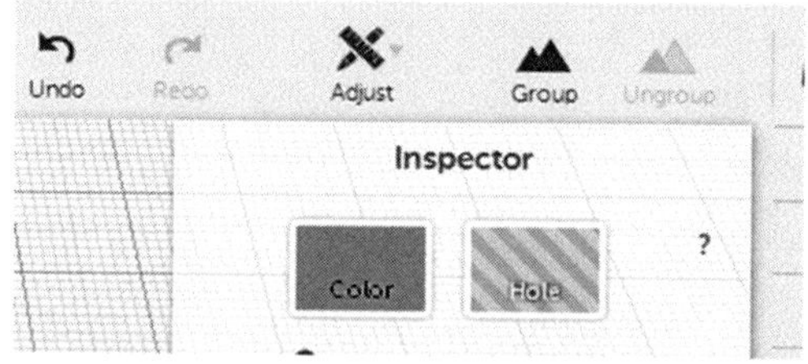

The Group feature lets the user combine holes and shapes to create custom shapes. Grouping also combines shapes to make one shape. To group multiple holes and shapes, click and drag a square around the shapes the user wants to combine, and then click the Group button. Note that it can take a few seconds for the grouping to occur after the user clicks the Group button. Below is an example of a box and a cylinder shape combined into a single object.

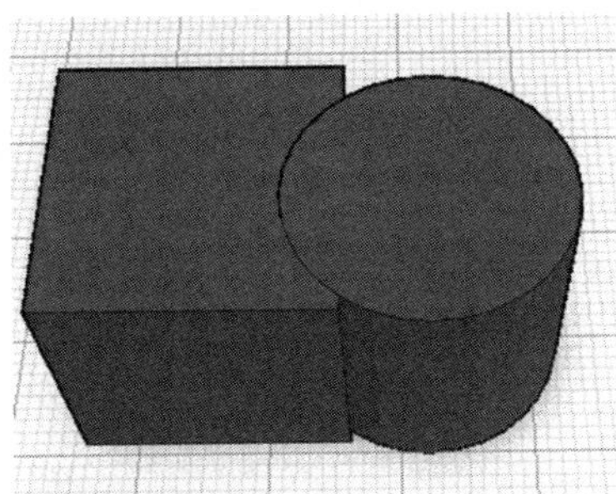

By using the Group tool, users can build objects by adding and grouping shapes multiple times. Many times, the same shapes are needed over and over. For example, an airplane needs two wings, and the user would want both wings to match. Once an object has been designed, the user can use the Copy and Paste feature in Tinkercad, which is very similar to the Copy and Paste feature in word processing programs.

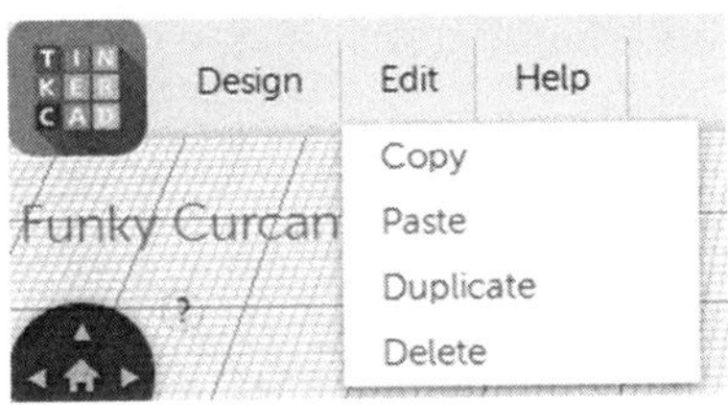

To copy a shape or object, click on the shape or object. Then click on Edit and Copy. Clicking on Edit and Paste will make an exact copy of the shape of object appear on the workplane. This can be done multiple times until the user has the number of matching objects desired.

Stacking Shapes

Shapes can be stacked on top of each other to make taller objects. To stack shapes, click on a shape that is on the workplane and locate the small black arrow at the top of the shape.

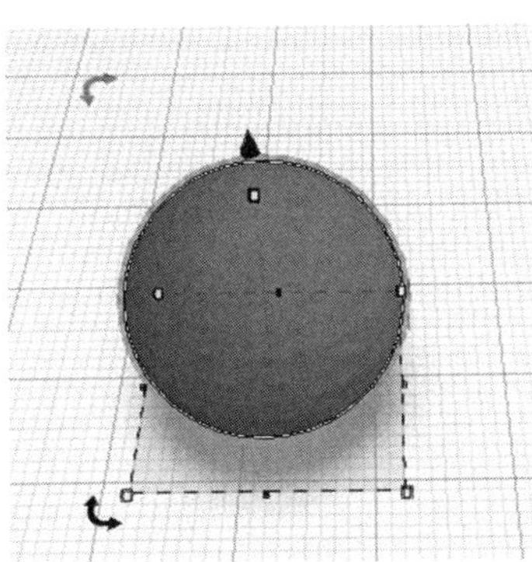

The small black arrow is the control that lets the user raise and lower the shape from the workplane. Clicking on the black arrow and dragging in an upwards motion will raise the shape from the workplane. The measurements shown reveal the distance the shape is from the workplane. A measurement of 4.00 mm means the shape is 4 mm above the workplane and is no longer resting on the workplane itself. Dragging the black arrow in a downwards motion will lower the shape. A negative measurement means the shape is below the workplane. A measurement of –4.00 mm means the shape is 4 mm below the workplane. Rotating the workplane will show that the shape is below the plane. To ensure shapes are sitting on the workplane, the black arrow measurement should show the measurement of 0.00, which means there is no distance between the shape and the workplane.

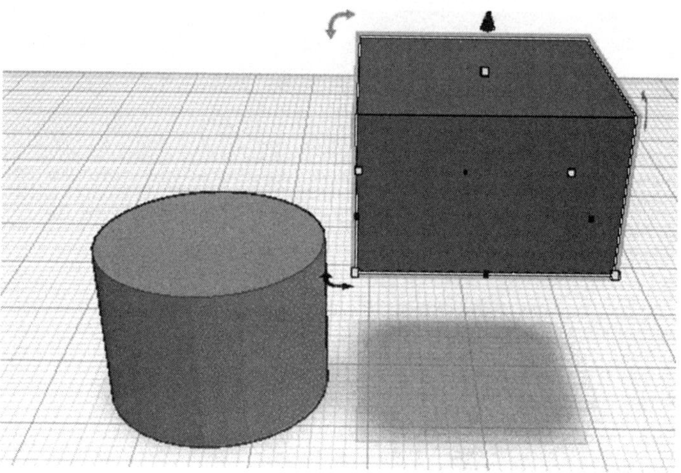

To stack shapes, click on the shape to be placed on top of another shape. Click on the black arrow and raise the shape until the measurement shows that it is above the shape that will be on bottom. After the shape is high enough, release the black arrow, click on the shape, and move it in place on top of the other shape. Once the shapes are stacked, they can be aligned so they are centered.

Aligning Shapes

The Align tool can be used to center shapes after they are grouped together. This creates an object that not only looks nice, but will result in a stronger object once it is 3D printed. To use the Align tool, click on the workplane and drag a selection square around the grouped shapes. Click on the Adjust button and then on the Align button.

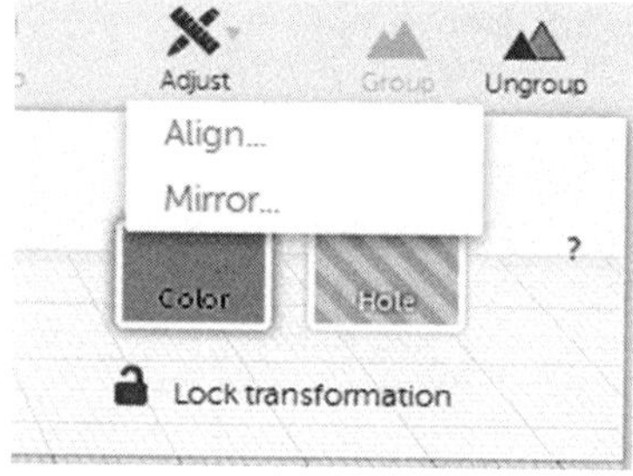

Gray alignment dots will show up around the object. Alignment dots for each face of the object will appear, and clicking on the alignment dot in the center of each face will center the shapes. Users can also align shapes along one side instead of centering the shapes by clicking the corresponding alignment dot on the shape face.

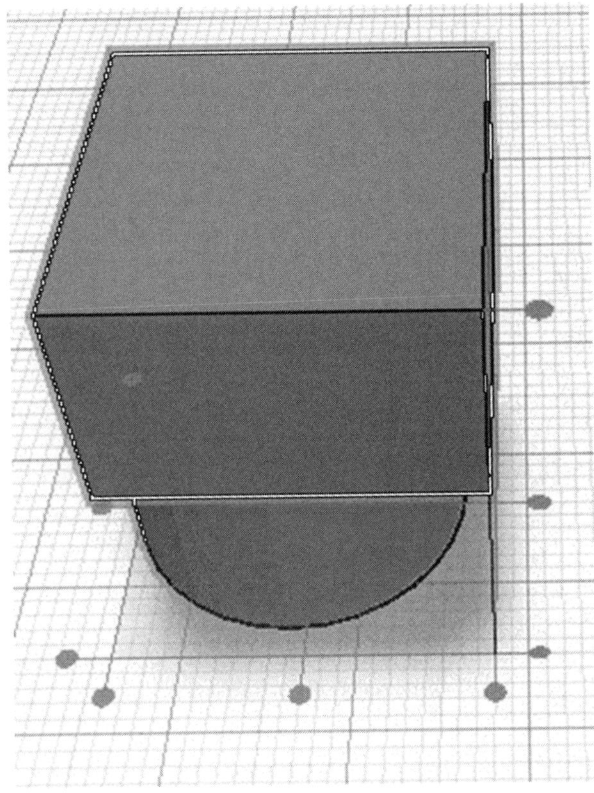

3D Printing Models Designed in Tinkercad

To print a model that has been designed in Tinkercad, click on the Dashboard and locate the model to be printed. The user will have the options of Download for 3D Printing, Download for Minecraft, and Order 3D Print.

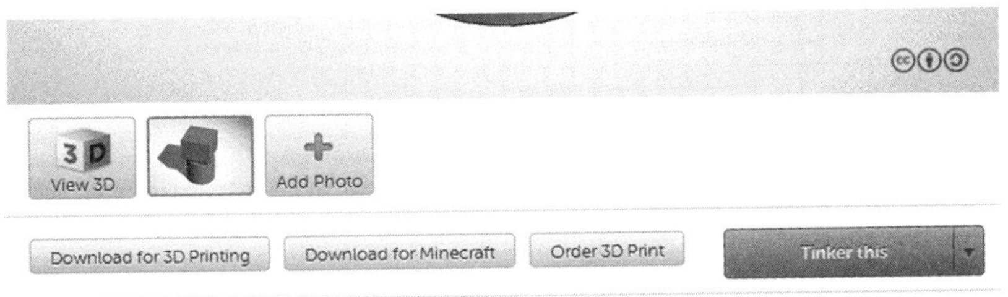

To 3D print the model, click on the Download for 3D Printing button. Users will have the option to download the model as .STL, .OBJ, .X3D Colors, or .VRML Colors.

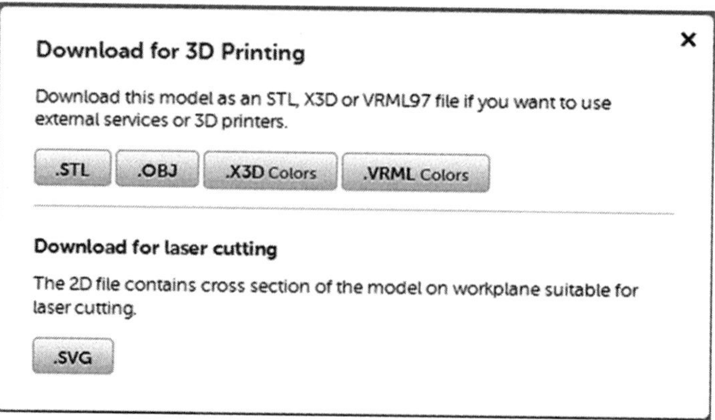

As described in Chapter 1, downloading model designs into .STL format will allow the models to be imported into other software programs, such as those used by individual 3D printers. Click on the .STL button and save the .STL file to the computer. Students should be taught how and where to save files to avoid confusion between students, especially when many models are being saved. .STL files of models can be saved to the computer desktop or to a flash drive. Either works just fine; the primary goal is to have a system for saving files and to teach that system to students.

The last step in 3D printing a model created in Tinkercad is to open the software used by the particular 3D printer being used and import the Tinkercad model. For example, if utilizing a Makerbot Replicator 3D printer, the user would open the Makerware software and then import the Tinkercad model. The Makerware program would convert the .STL file from Tinkercad into a design that can be printed using the Makerbot.

Tinkercad and Minecraft

The figure below shows other options for Tinkercad models. One of the options is to download the designed model into Minecraft.

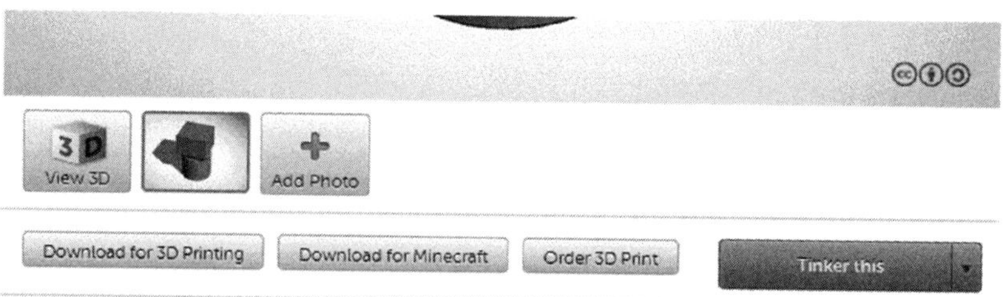

Minecraft is a video game that can be played on computers (minecraft.net) or through an app on a tablet. Minecraft is a very popular game that is played by students of all ages. Many schools and libraries are offering Minecraft clubs for students, and integrating 3D modeling would be a great extension to a program.

Choosing the Download to Minecraft option will convert the model into an object that can be placed inside the game player's virtual Minecraft world. Users have the opportunity to size the object according to Minecraft specifications.

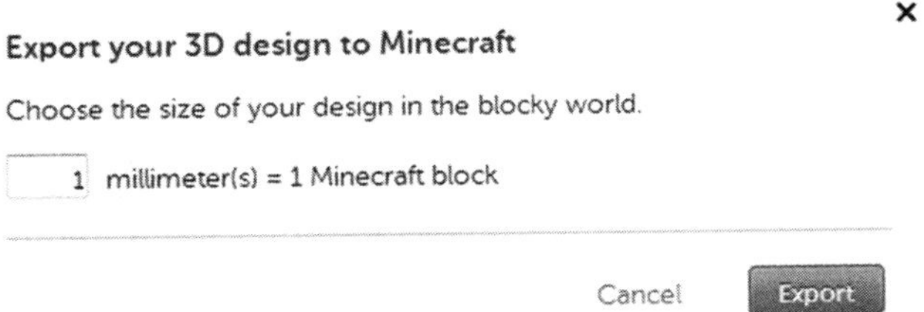

Users will need to download MCEdit from mcedit.net. MCEdit is the program that allows users to import their models into Minecraft. Click Import and select the location where the model should be placed in Minecraft. Click the Import Lock button and then Save. Quit MCEdit and then open Minecraft to see the model that was imported.

Order 3D Prints

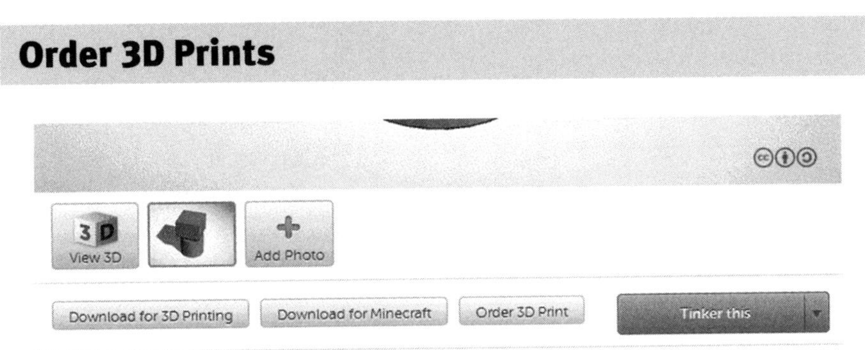

Even if the library does not have a 3D printer, students can still benefit from the 3D modeling process. Clicking on the Order 3D prints button allows users to choose a 3D printing service that will print the model for a fee and send it through the mail. Tinkercad partners with Ponoko, Sculpteo, Shapeways, and i.materialize to print models. These sites allow users to choose the material in which to print the object. The larger the object or the more valuable the print material, the higher the cost will be for the user.

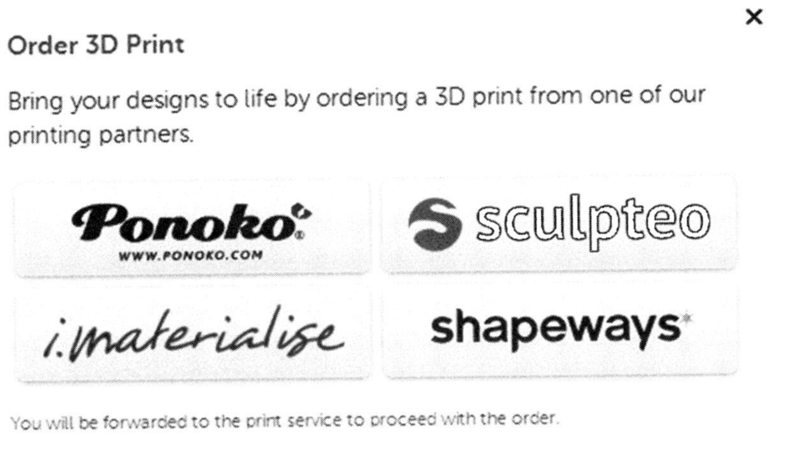

Tinkercad for More Experienced Users

Although Tinkercad is simple to use, even for students in elementary school, it does have some features for more experienced designers. One of these features allows users to import their own designs. To import a design, select Import from the list on the right side of the workplane.

- Favorites
- Import
- Shape Generators
- Helpers
- Geometric
- Holes
- Letters
- Number
- Symbols
- Extras

Designs can be imported from the user's computer or from a website on the Internet. This feature allows users to import models from sites such as Thingiverse. If students use Thingiverse as a resource to find 3D models, they should be taught to read the license information for the model before they download and import it. The license terms tell users what is and is not permitted with that particular model. This is a good way for librarians to teach good digital citizenship and for students to learn that just because something is on the Internet, that doesn't mean the design is available to use for any purpose. If students copy a model in Tinkercad or Thingiverse that another user designed, students should be taught to always give credit to the original designer when they make a copy.

Chapter 4 Key Points

Tinkercad can be accessed at Tinkercad.com.

Tinkercad is a good beginning program for students who have had absolutely no experience in digital modeling.

Skills learned by using Tinkercad can be transferred to most other 3D modeling software programs.

Tinkercad works well using Google Chrome 10 or Mozilla Firefox 4 (or newer) browsers.

If accessing Tinkercad on a Mac using Safari, WebGL has to be enabled before accessing the website.

Tinkercad users are able to add a photo and information to their profiles, but librarians should be cautious. Students should *not* add photos of themselves or personal information to their account profiles.

After logging in for the first time, users will see four step-by-step lessons that can be used to get to know Tinkercad and how it works.

Tinkercad has three menus that allow users to easily access models and designs: the Dashboard, Gallery, and Learn menus.

5

Focus on 3D Modeling: 123D Design

123dapp.com/design

Autodesk is a company with multiple 3D modeling apps and programs. Autodesk owns Tinkercad, as well a variety of other 3D modeling programs and apps. One of these 3D modeling programs is 123D Design. Like Tinkercad, 123D Design can be accessed on the Internet and is fairly easy for beginners to use. 123D Design uses WebGL, so Google Chrome and Mozilla Firefox are both good browser choices for 3D modeling design using 123D Design. If users have difficulty with the WebGL capabilities of either Google Chrome or Mozilla Firefox, the other browser can be used to access 123D Design. Students can be taught that if, for some reason, one of these browsers doesn't work, they can problem solve by trying the other browser.

To begin using 123D Design, go to 123dapp.com/design.

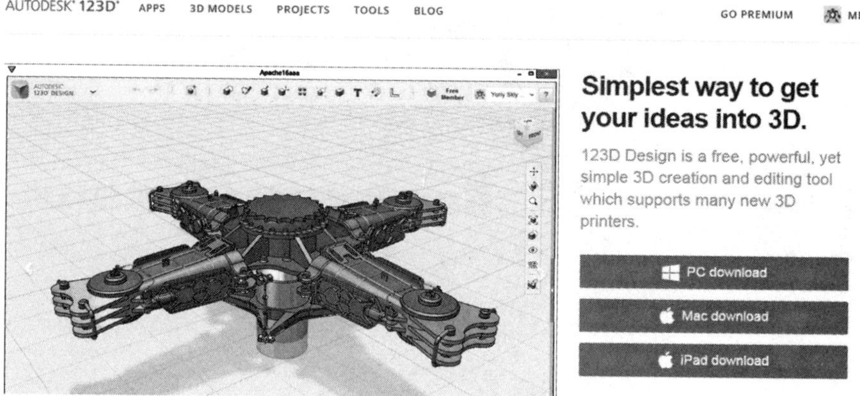

Users have the option of downloading 123D Design or launching 123D Design online. Launching the program online is a good option because users do not have to go through the downloading process and can access the program wherever there is an Internet connection. Downloading the program will install 123D Design onto the computer, and users would not need an Internet connection to use it. Because both options work equally well, users can choose the option that works best for their situation.

123D Design offers free and premium accounts. Free accounts are for non-commercial use of 3D models, and users can design unlimited free 3D models and up to ten premium 3D models per month. After clicking on a premium model to move to the design grid, the user will see a message that the model is a premium model, and the user can choose whether or not to proceed with moving the model to the grid. The free accounts are more than adequate for most students new to 3D modeling and printing. Creating a free account is easy; enter an e-mail address and password, and the free account is established.

123D Design has a simple layout similar to Tinkercad but has more features that students will enjoy learning and growing into.

Toolbars and Menus

The site opens to a design grid with toolbars above and to the right of the design grid. The main toolbar runs horizontally along the top of the screen. This toolbar contains Save, Undo, Redo, Copy & Paste, Switch Active Planes, Precise Move, Precise Scale, Precise Rotate, Ruler, Finishing Tools, Grouping, and Text buttons.

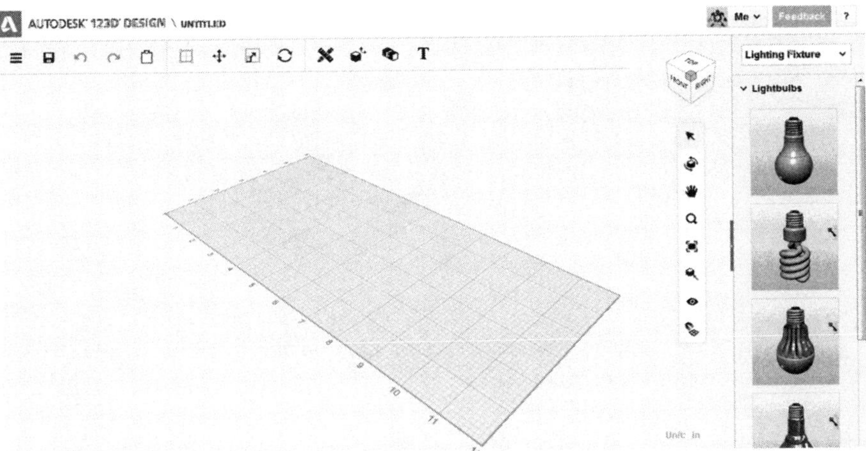

The navigation toolbar is a vertical toolbar to the right of the design grid.

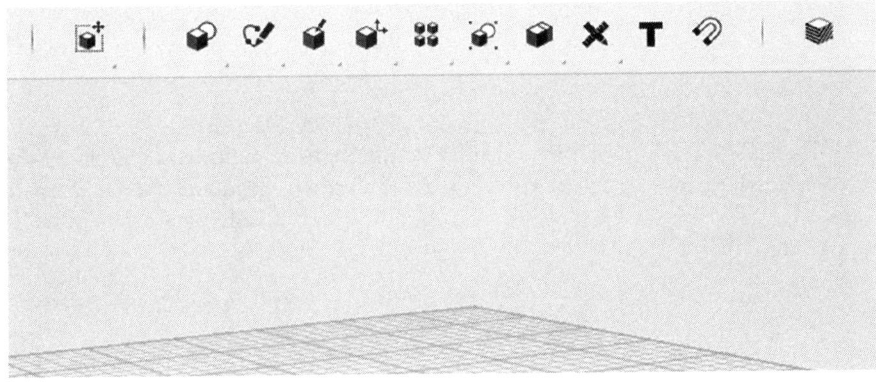

This toolbar has buttons for:

Orbit – rotates the view

Pan – moves camera from side-to-side

Zoom – zooms to cursor location

Zoom to Fit – zooms to fit the view

Look at – to look at selected face

Orthographic Mode – this mode gives a different view and makes aligning objects easier. Users can toggle between seeing just materials, outlines, or both materials and outlines on the design grid.

Grid Settings – changes grid measurements to inches, centimeters, and millimeters and can change grid size. Users can also set up the grid size to match the size of the print plate of the 3D printer being used.

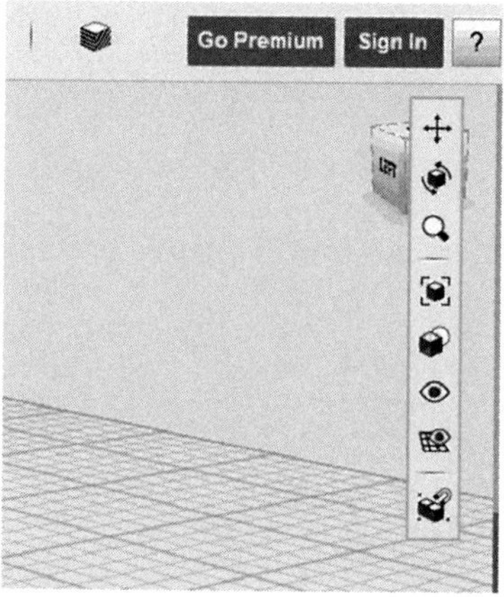

To access the main menu, users should click on the three horizontal bars in the top left corner on the main toolbar.

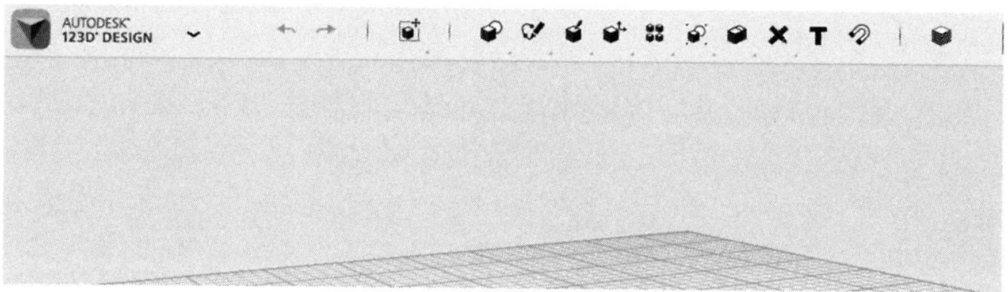

The 123D Design website includes comprehensive help by clicking on the Question Mark button at the top right of the screen. Students should learn to utilize help features when they get stuck as a first step in solving a problem while designing.

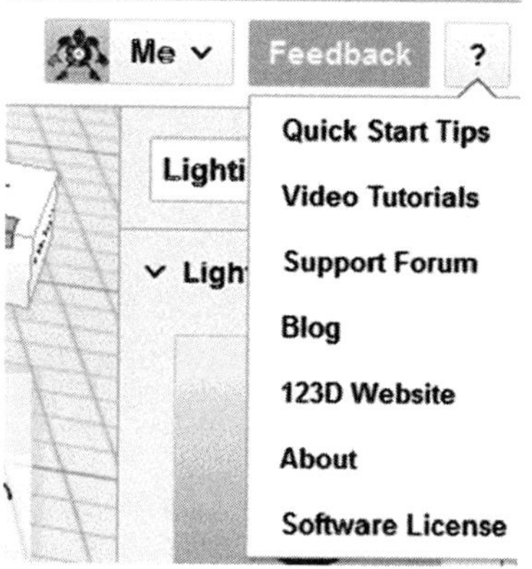

Users can access video tutorials, quick tips, the 123D Design blog, and a user forum from the help button. Autodesk and 123D Design also have their own YouTube channel, which makes finding and viewing video tutorials easy. The user forum is a great place to ask questions and receive answers or to search questions and responses by other 123D Design users.

Designing 3D Models in 123D Design

To begin a design, open the list of shapes on the right side that can be selected and placed on the design grid.

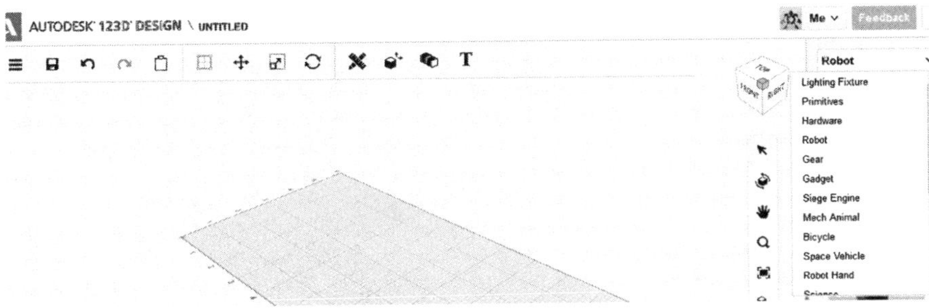

To see the list of options, click on the pull-down list on the right side of the design grid. To find basic shapes, locate Primitives to the right of the grid and click to open.

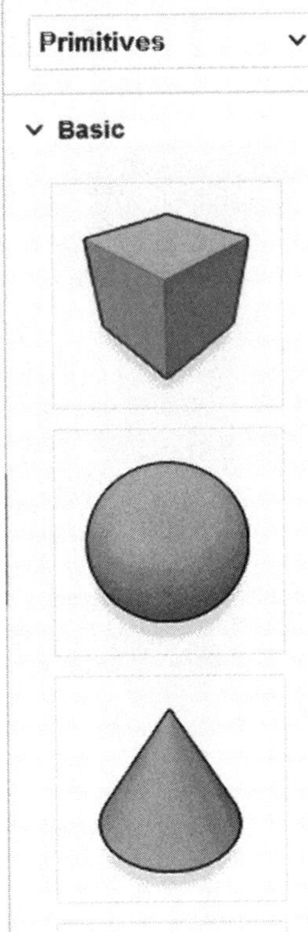

Shapes will appear in a list on the right side of the screen. To move shapes to the design grid, click on the shape and drag it to the grid. The View Cube can be used to change the view of any object on the grid.

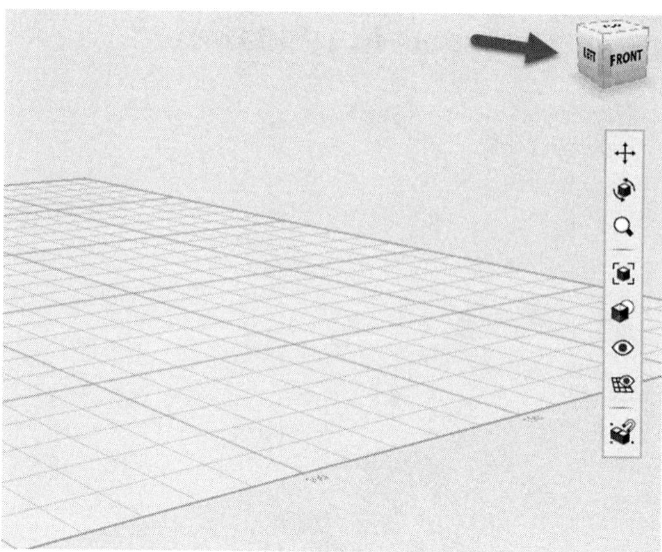

Changing Dimensions of Objects

Clicking on a shape or object on the design grid will show the manipulator controls. These controls look very similar to the controls in Tinkercad and work in a comparable manner.

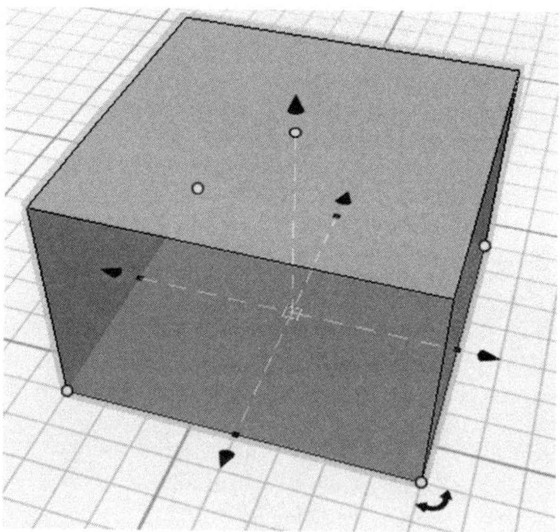

The manipulator controls include white circle grips on the corners and small black square grips at the midpoints. Clicking and dragging on the grips change the dimensions of the object on the grid. Clicking and dragging the rotation arrow rotates the object on the design grid. Dragging the rotation arrow rotates the object around its center. The black cone-shaped grips allow the user to move the object on the design grid without changing its dimensions.

Clicking on an object on the design grid triggers the Attributes bar. Double-clicking on an object displays the Paint Palette square. Clicking on this will show the numerous colors that can be used for objects on the design grid.

Attributes

The attributes bar shows the length, width, and height of the object on the design grid. Users can enter any dimension values in the length, width, and height boxes to easily change the dimensions of the object at any time during the design process.

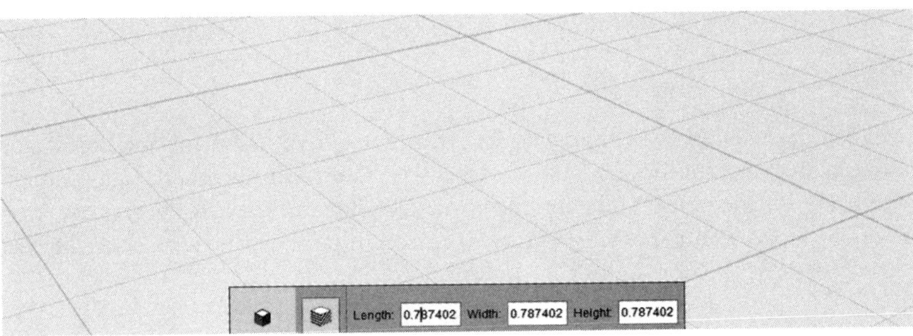

Grouping Shapes and Objects

To group shapes or objects together, first drag two shapes to the design grid. To stack shapes, simply drag one of the shapes on top of the other and the shapes will snap together. Clicking on each individual shape will allow the dimensions for that shape to be changed or the shape to be scaled. Shapes and objects can also be grouped by clicking on the grid and dragging a square around all of the objects to be grouped. Right-clicking and selecting Group will group those objects together.

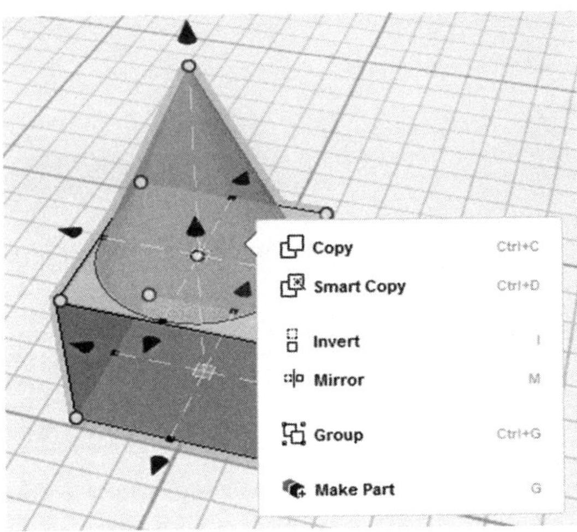

Objects can be solid or other shapes can be inserted into a solid and turned into a void. This creates a hollow in the solid shape.

Exporting Designs to .STL

Unlike Tinkercad, designs cannot be directly exported to .STL file format from 123D Design, but designs can be downloaded and converted to .STL fairly easily. To download a design, click on the Save button in the top menu bar.

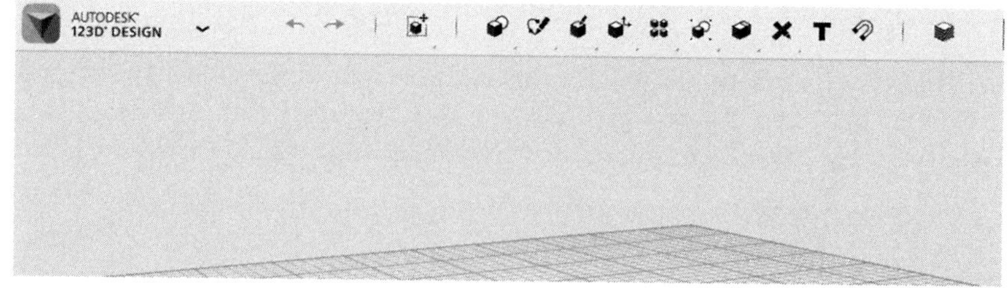

The Save window will appear. Enter a descriptive name for the object and choose to make the design public or private. Users can also add tags to make searching for the object and a description easier, but these are not necessary. After entering a name and choosing to make the design public or private, click the blue Save button at the bottom of the window.

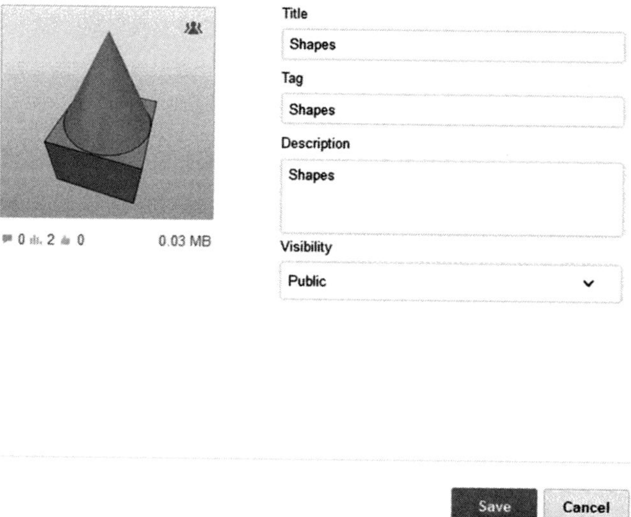

Next, click on the profile avatar in the top right corner of the screen.

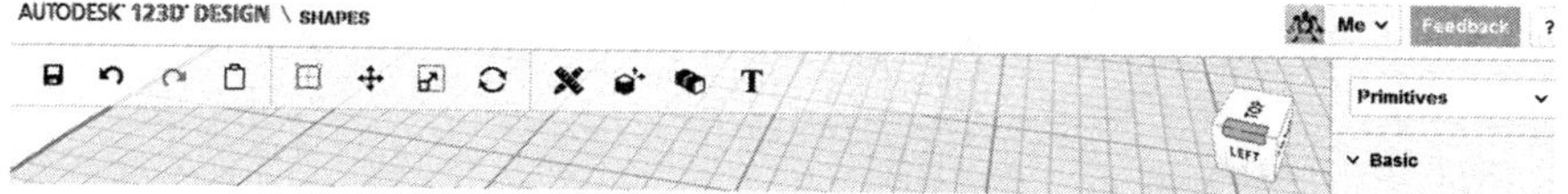

Click on the My Projects button. This will open a new window, and the model or project to be downloaded will appear. If the design does not appear, click on the model tab on the left side.

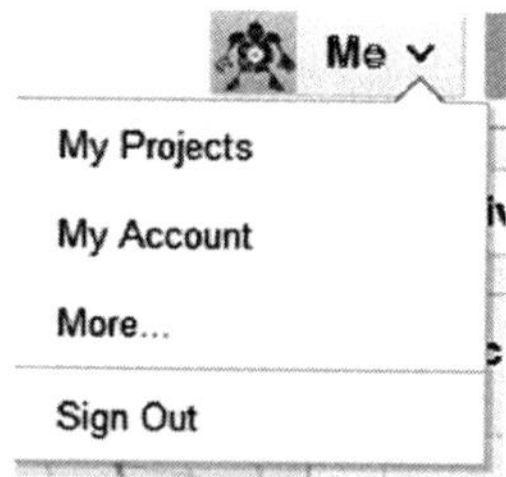

Click on the picture of the model to be downloaded and then click on the blue Edit/Download button.

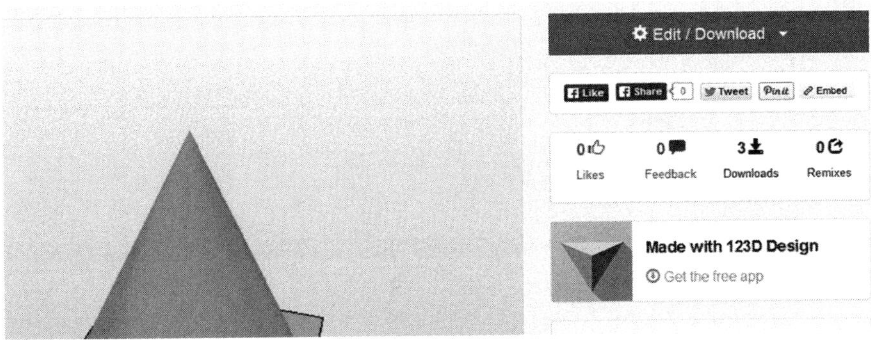

Scroll down and click on Download 3D Models.

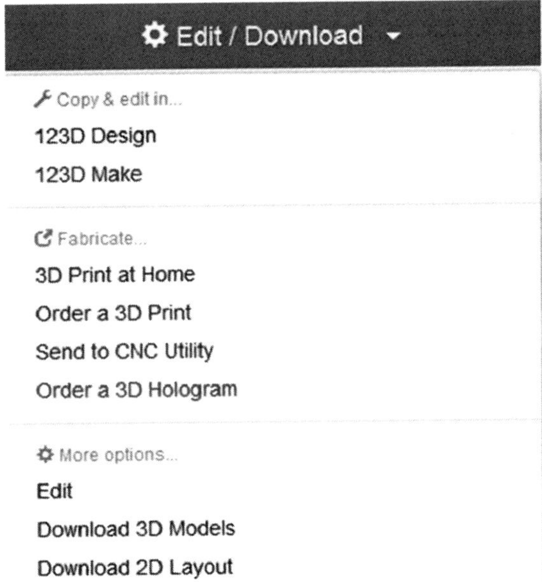

Uncheck all of the boxes EXCEPT the box next to the .STL file. Then, click on the blue Download Models button.

The design can now be saved to the desktop of the computer. The program for the specific 3D printer being used can be opened and the saved design can be imported into that program for printing.

123D Design is also available as a free app in the iTunes store for IOS devices. The app has many of the same capabilities as the web version and is a good app for beginners to 3D modeling.

Chapter 5 Key Points

123D Design can be accessed at 123dapp.com/design.

123D Design can be downloaded to a computer or launched from the 123D Design website.

Launching the program online is a good option because users do not have to go through the downloading process and can access the program wherever there is an Internet connection.

Downloading the program will install 123D Design onto the computer, and users would not need an Internet connection to use it.

To access the main menu, users should click on the three horizontal bars in the top left corner on the main toolbar.

The 123D Design website includes comprehensive help by clicking on the Question Mark button at the top right of the screen.

Clicking on a shape or object on the design grid will show the manipulator controls.

The user can enter any dimension values in the length, width, and height boxes in the Attributes bar to easily change the dimensions of the object at any time during the design process.

6

Focus on 3D Modeling: 3DTin

3Dtin.com

3D Tin is a browser-based program that works in the Google Chrome and Mozilla Firefox browsers because of their compatibility with WebGL protocol. This program stores models in the cloud, which makes them accessible from anywhere. 3DTin offers video tutorials to help beginners learn the 3DTin interface. 3DTin is easy enough for students who are brand new to 3D modeling. Users create models in 3DTin by building with blocks and adding geometric shapes.

To access 3DTin, go to 3DTin.com. 3DTin opens to an empty design grid. Like other modeling programs, the design grid is where the modeling design takes place.

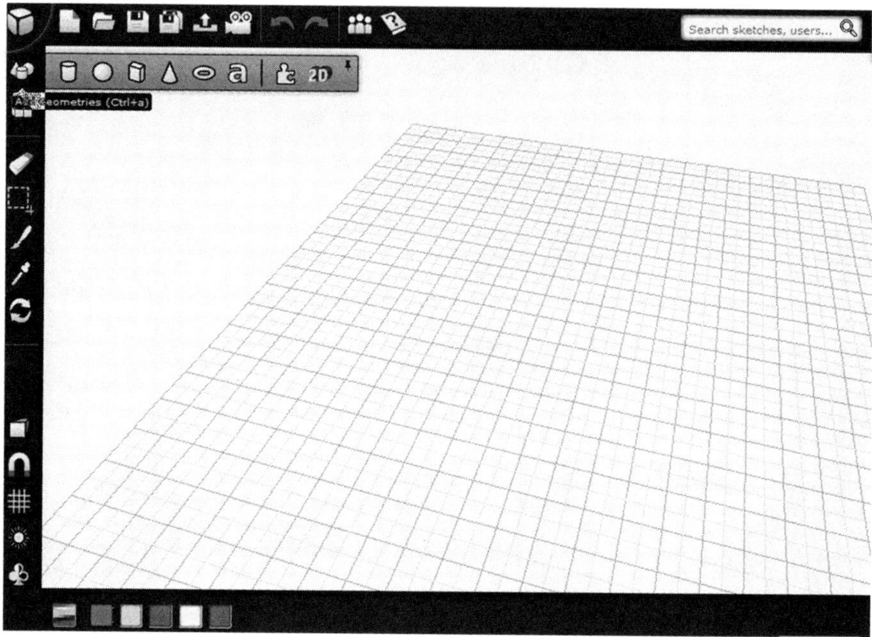

Tip: Mouse over buttons around the grid and cues pop up that describe in a word or two what each button does.

Designing a Model

To begin a design in 3DTin, click on the Add Cubes button, which is on the toolbar on the left side of the grid.

Clicking on the Add Cubes button turns the cursor into a transparent cube. Move the cursor to a location on the design grid and click. The cube is then placed at that location on the design grid.

To create a line of cubes, click and drag the cube. To erase one or more cubes, click on the Eraser tool in the toolbar and then click on the cube(s) to be erased.

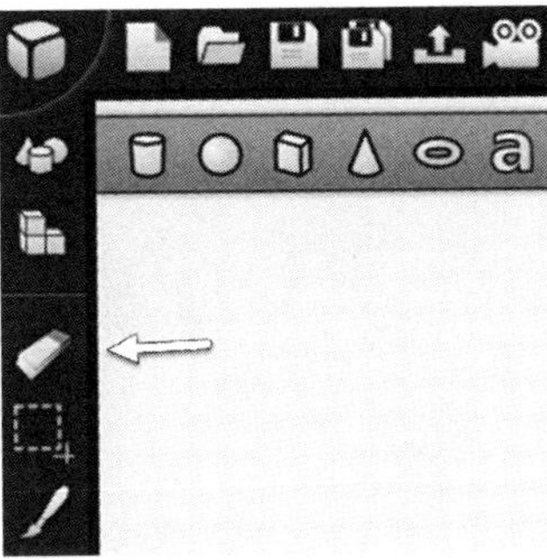

Cube colors can easily be changed by clicking the color bar below the design grid, clicking on the Paintbrush button on the toolbar, and then clicking on the cube to change the color.

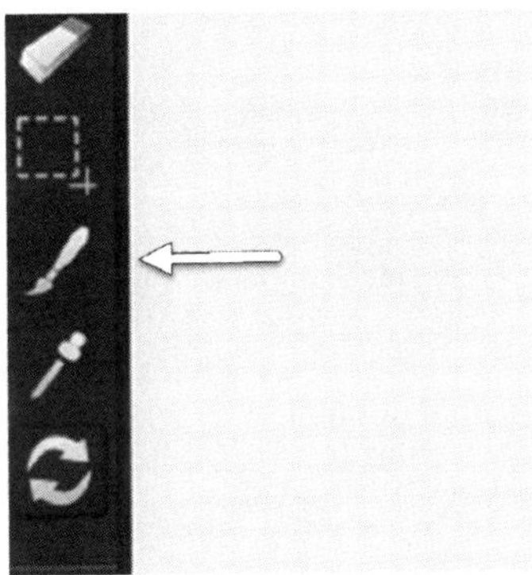

Users can augment cube designs by adding geometric designs. To do this, click on the Geometries button.

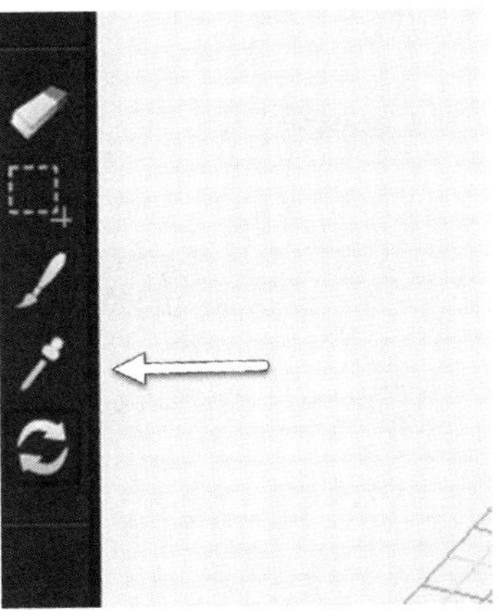

Clicking the Geometries button will open the Geometry Builder box with over fifty shapes that can be added to the design grid.

Options for adding words to the design grid are located at the bottom of the Geometry Builder. Users can add text or a text void. Choosing the Text option will add a solid to the grid. Selecting Text will allow the user to customize the text to be added to the design grid, designate the width, height, and thickness of the text, and select a font for the text. Users can choose from the three fonts provided. Choosing the Text Void option will generate a void in the shape of text. The Text Void creates an engraving of text into a solid shape.

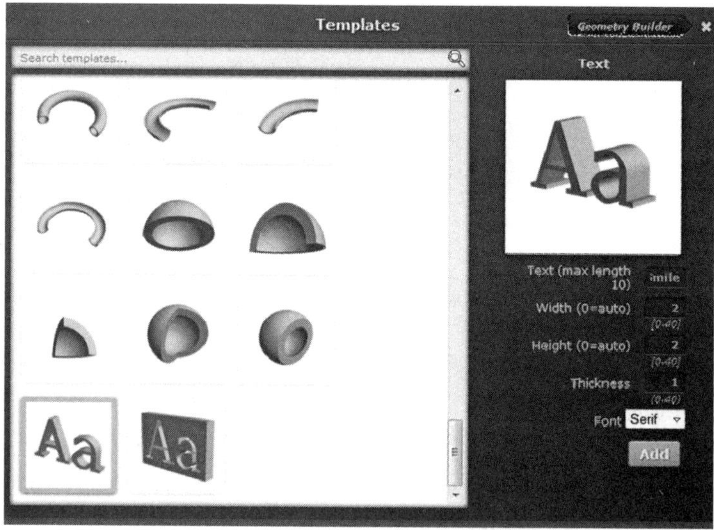

Clicking the green Add button will show the solid shape or text on the design grid. Clicking anywhere on the grid will drop the shape or text onto the grid. An exact copy of the shape or text can be added by clicking again. This can be done any number of times to place multiple copies of the same shape onto the grid. After adding shapes to the design grid, click on the View Rotate (or Pan) button.

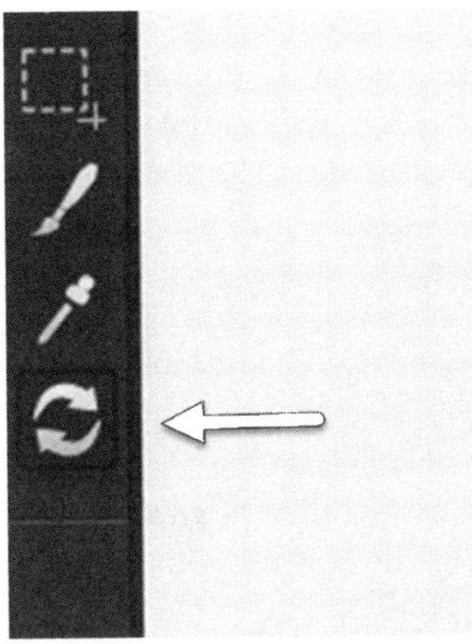

Templates

The template toolbar is a quick way to access shape or text templates.

The template toolbar provides quick access to cylindrical, spherical, rectangular, conical, toroidal, and text templates. This toolbar also allows users to build custom geometries by subtracting a set of solids from other shapes and clicking the Make Your Own Geometry button. Clicking this button will launch the Geometry Builder, which allows users to add primitive geometries. After adding the shapes, click on them to change their properties and apply different transformations. If the Subtract box is checked, the geometry will be highlighted in red, and in the preview section, the shape will be subtracted from the other solid (3DTin Blog 2013).

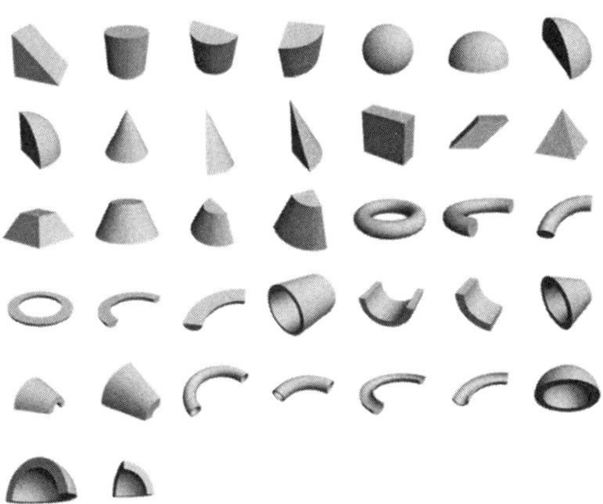

When the shape is complete, click the Build button. The shape will be baked and can then be added to the model. Users can go through this process any number of times. To make any changes to it in the future, click on it and press the icon. The geometry will be opened in the Geometry Builder and will look just as you left it.

Designing Shapes with Curves

To create a shape with a curve, click on the 2D button on the Shapes bar to launch the 2D editor. The 2D editor presents you with a smooth curve along with a bunch of control points. By moving the control points around you can change the shape of this curve (3DTin Blog 2013).

There are two ways in which this curve can be used to build a solid: Extrusion and Revolution.

Extrusion

Clicking Build will automatically close the curve and extrude along the third dimension, and a solid of specified thickness is ready to add to your main sketch. The number of control points can be increased up to 30, which allows users to add more detail to the curve created (3DTin Blog 2013).

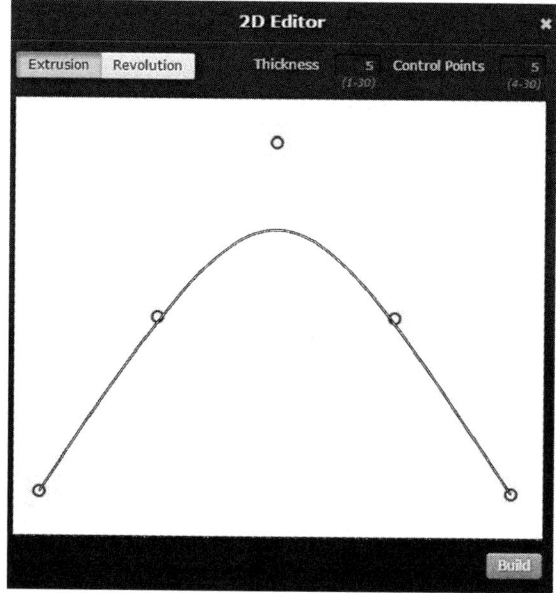

Revolution

Clicking Build causes the curve to revolve around the shown axis of revolution, and a final solid is presented to add to your sketch. This option works well when creating objects that are symmetric, such as vases.

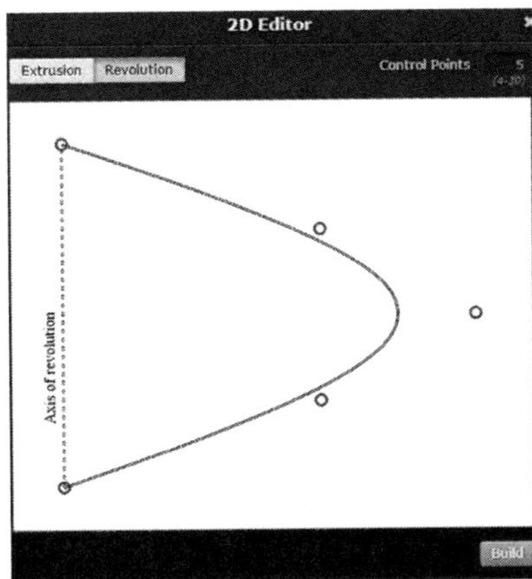

Orthographic View

Orthographic view is a way of drawing 3D objects that is primarily used by architects and designers. The key difference between Orthographic and Perspective projection is the change in the size of objects depending on their distance from the viewer. In orthographic projection, as an object moves away from the viewer, its size doesn't appear to change; in perspective projection, however, it appears to get smaller.

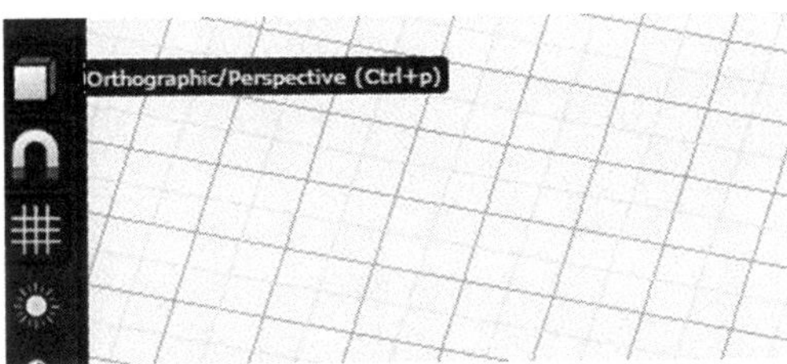

The view can be changed from Orthographic to Perspective by clicking on the Orthographic/Perspective button on the left side.

Snap to Grid

The Snap to Grid button is on the bottom left toolbar.

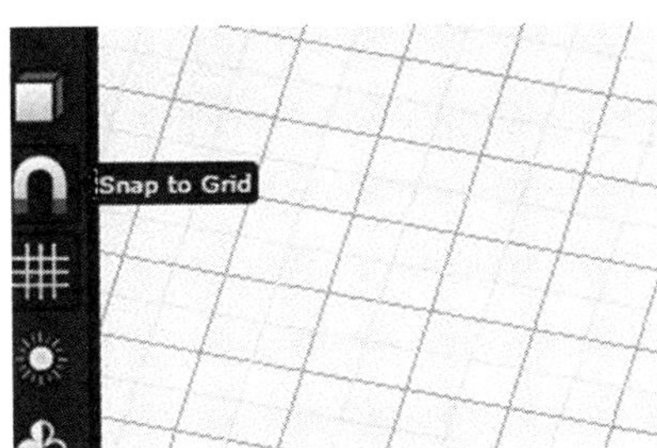

When the Snap to Grid feature is turned on, shapes will snap to the grid lines as the user moves them around. When this feature is turned off, shapes do not snap to grid lines, and the user is able to move them in a continuous trajectory. The position panel at the bottom continuously displays the location of the shape on the grid.

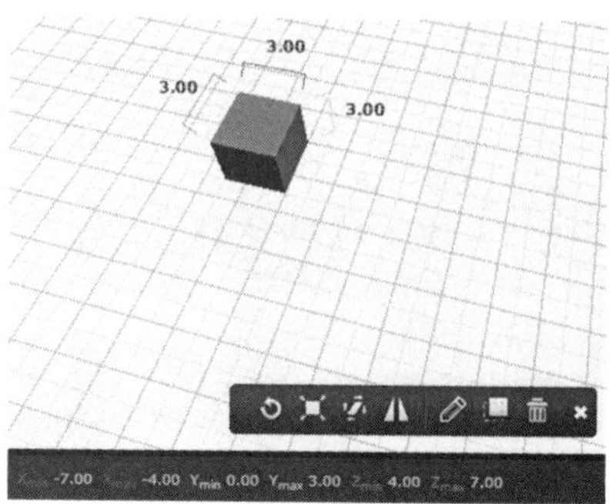

Grouping

To group shapes together, click on the Select icon and draw a rectangle over the design. This will capture all shapes to be grouped.

Once the shapes to be grouped have been selected, a toolbar will appear at the top. Click the Group button and the individual shapes will become one object. This newly created object can now be moved, rotated, and flipped.

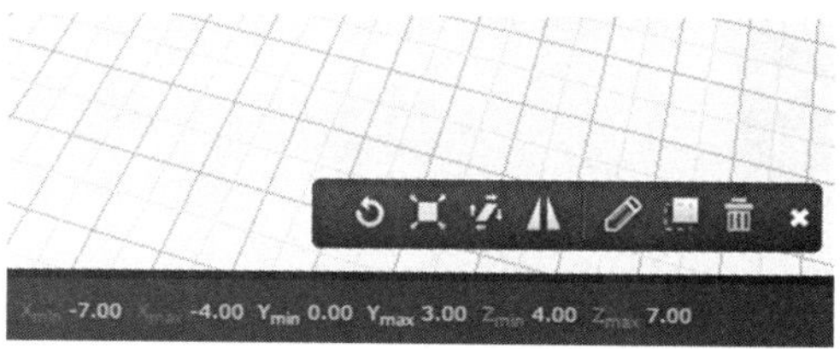

Exporting Models

Save the model and click on the Export button in the top toolbar. Users will then have the option of exporting the file in OBJ or STL format.

An OBJ file downloads as a zip archive, which contains .obj and .mtl files. The STL file is in a binary STL format that can be used with 3D printing services.

Community

Clicking the Community button will open a new window with tabs for Group Chat and Forum.

Group Chat

This is a realtime chat application for 3DTin users who are logged in. Users can post questions, share 3D modeling ideas, and help others by answering questions. The chat application has built-in notifications that let users know when others are actively chatting. This feature can also be turned off so users don't receive notifications if they don't want to be interrupted.

While this feature could be useful for more experienced modelers, librarians should be cautious if allowing students to use this feature. This would be a good time to review digital citizenship guidelines with students to make sure this feature is used productively and safely.

Forum

Another option for connecting with other users is the Forum. Clicking this tab allows users to post questions to the 3DTin Google Group. This provides users a more traditional mailing list way of exchanging information.

Chapter 6 Key Points

To access 3DTin, go to 3DTin.com.

3DTin is a browser-based program that works in Google Chrome or Mozilla Firefox.

3DTin offers users video tutorials to help beginners learn the 3DTin interface.

When users mouse over buttons around the grid, cues pop up that describe in a word or two what each button does.

To begin a design in 3DTin, click on the Add Cubes button, which is on the toolbar on the left side of the grid.

Users can add text or a text void to their model. Choosing the Text option will add a solid to the grid. Choosing the Text Void option will generate a void in the shape of text.

The template toolbar is a quick way to access shape or text templates.

To export the model for printing, save the model and click on the Export button in the top toolbar. Users will then have the option of exporting the file in OBJ or STL format.

7

Focus on 3D Modeling: SketchUp

SketchUp.com

SketchUp Make or SketchUp Pro?

SketchUp is a 3D modeling program that can be downloaded from the SketchUp website at sketchup.com. There are two versions: SketchUp Pro, a paid version, and SketchUp Make, a free version that has more than enough modeling capabilities for students. If students move beyond the features of the free version, the company offers free Educator licenses. To receive an Educator license for SketchUp Pro for free, the educator must teach at an accredited educational institution and be able to provide a current faculty identification card. The Educator license, which can be installed on both laptops and desktops using either Windows or Mac OS X, expires annually and can be upgraded to new versions for free. This license cannot be used for any commercial purposes, and should only be used for student learning.

For most students, beginning with SketchUp Make should be adequate, unless the students have prior experience with 3D modeling. SketchUp Make allows users to design for free, as long as the designs are for noncommercial purposes. SketchUp may have slightly more of a learning curve than other 3D modeling programs such as Tinkercad, but is still a fairly easy modeling program to learn, especially for older students. As with other 3D modeling programs, users can find many tutorials on YouTube to help get started.

Getting Started with SketchUp

To begin, download SketchUp Make from the SketchUp website (sketchup.com) and click on the SketchUp icon to open it up.

The first time that SketchUp is opened, a Welcome to SketchUp window will appear.

This window will also allow users to access SketchUp templates. Click on the Choose Template button in the top right corner. To begin, select the top choice, Simple Template - Feet and Inches, and then click on the Start Using SketchUp button in the lower right corner of the window. If the Welcome window doesn't appear, this dialog box can be accessed at any time from the menu bar.

Modeling Axes

SketchUp opens to the modeling window that has a toolbar along the top and red, blue, and green lines in the workspace. The red, blue, and green lines represent the x, y, and z modeling axes and help the user gain a sense of direction during the design process. When a line or object is drawn or moved parallel to a line, the model is being built in that direction. To move an object up or down, the user would move the object parallel to the blue line. To move an object to the back and front, the user would move the object parallel to the green line. Moving an object parallel to the red line will move the object to the left or right of the design plane. The colored axis lines do not show up when the design is exported.

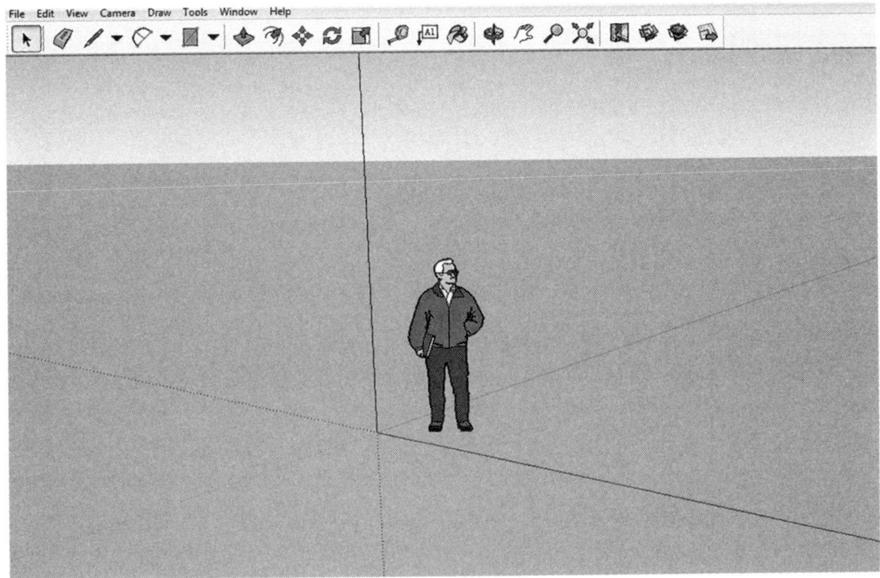

The Getting Started Toolbar

SketchUp has several toolbars, but the one on the main window contains many of the tools needed for basic modeling. The Getting Started toolbar contains most, but not all, of the tools needed to design a variety of 3D models. Tools on the Getting Started toolbar include the following:

- Select – select to change objects using other tools and commands
- Eraser – remove or soften objects
- Lines – changes between Line and Freehand
- Arcs – changes between 2 point arc, arc, or pie
- Shapes – changes between rectangle, circle, or polygon
- Push/Pull – push and pull faces to sculpt objects
- Offset – offset edges in a plane
- Move – move, stretch, and copy objects
- Rotate – rotate, stretch, and copy objects around an axis
- Scale – scale and stretch objects

- Tape Measure Tool – measure distances and scale objects
- Text – add text
- Paint Bucket – apply color and material to an object
- Orbit – orbit camera around the model
- Pan – pan the camera vertically and horizontally
- Zoom – zoom the camera in and out
- Zoom Extents – zoom the camera in and out to show a whole model
- Add Location – add a geo-location to a model
- Get Models – find models in 3D Warehouse
- Extension Warehouse – add extensions
- LayOut – send to LayOut (in SketchUp Pro)

Toolbars

Other toolbars can be viewed by clicking on View and then Toolbars. Users can view the Advanced Camera Tools, Camera, Classifier, Construction, Drawing, Edit, Google, Large Tool Set, Layers, Measurement, Outer Shell, Principal, Sandbox, Sections, Shadows, Standards, Styles, Views, and Warehouse toolbars by clicking View, then Toolbars, and placing a checkmark next to the toolbar to be displayed.

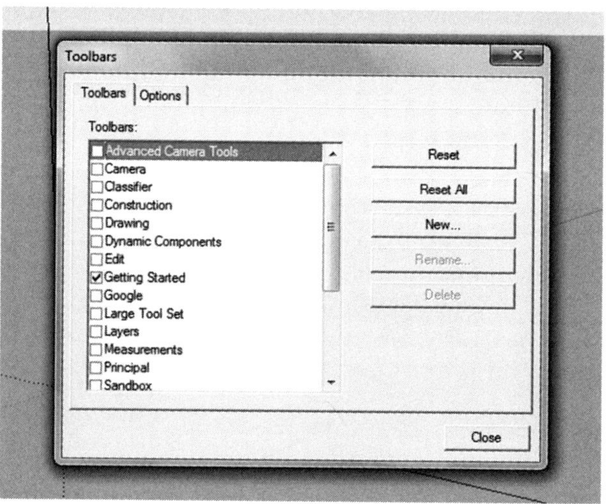

Status Bar

Along the bottom of the design screen is the Status Bar.

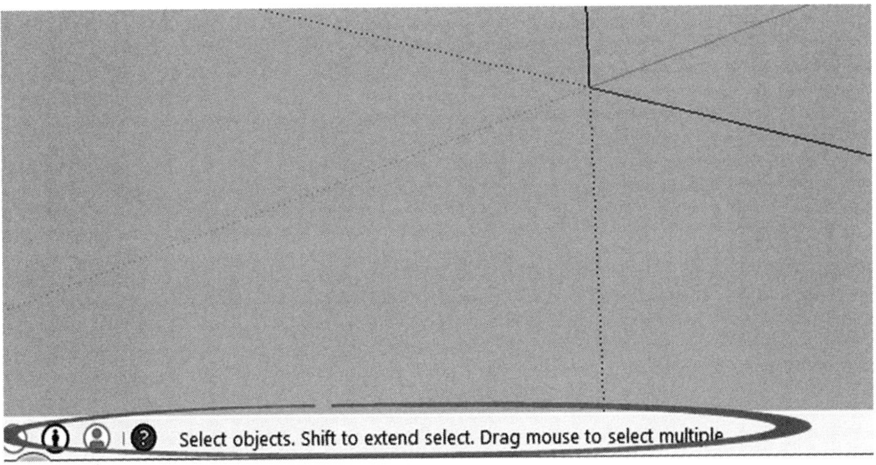

The Status Bar shows whether the model is geo-located, shows credits for designs if the model has been imported from 3D Warehouse, and provides quick access to the Instructor, which is the SketchUp help menu. Clicking on Advanced Options on the Instructor will take the user to the online knowledge center. The online knowledge center is a site with much information on using SketchUp. The Status Bar also provides users with instructions on the tools that are selected for use.

Measurements Box

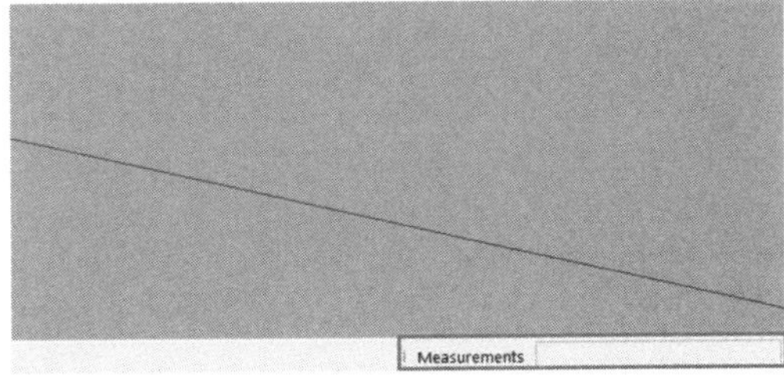

To the right of the Status Bar, in the lower right corner of the screen, is the Measurements box, one of the most important tools to be aware of while designing a model.

The Measurements box shows the user measurements of whatever tool the user is applying. The Measurements box will show the length of the line or the radius of the circle being drawn. Users can also enter measurements in the Measurements box and an object of that size or length will be automatically created.

3D Modeling in SketchUp

To begin a new design, first clear the 2D figure of the person shown on the design plane. This figure always appears when a new plane is opened. SketchUp changes the person represented on the plane on a regular basis.

To get rid of the figure, click on the Eraser tool in the Getting Started toolbar and then click on the figure, and it will disappear.

To add one of the basic shapes to the design workspace, click on the Shapes button on the Getting Started toolbar.

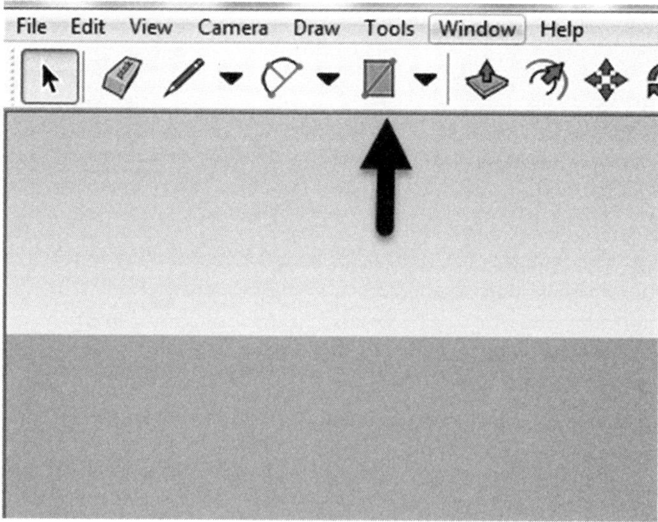

Clicking on the workspace will drop one side of the shape on the workspace, and clicking again will drop the other side of the shape.

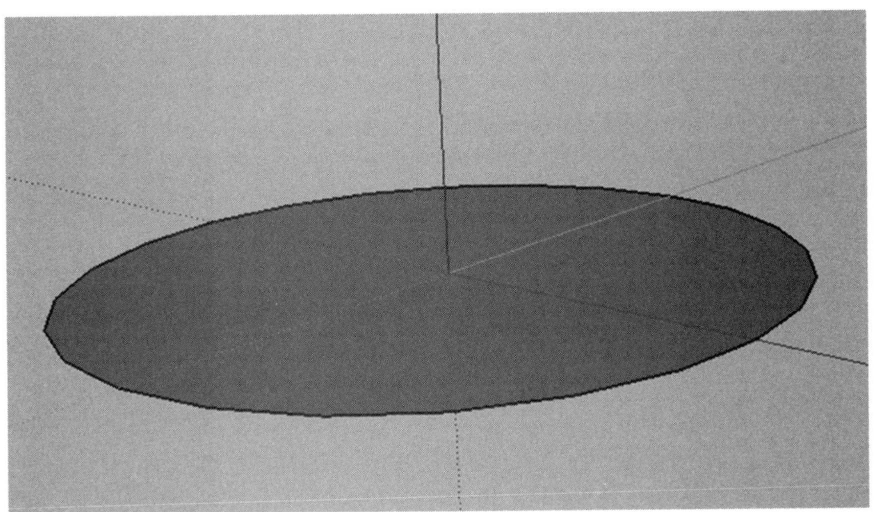

The Push/Pull feature can be used to make this 2D shape into a 3D figure.

To do this, click on the Push/Pull button and then click a face of the shape and drag the cursor. Clicking again will stop the Push/Pull feature. The Push/Pull tool only works on flat faces and will not work if the face is curved.

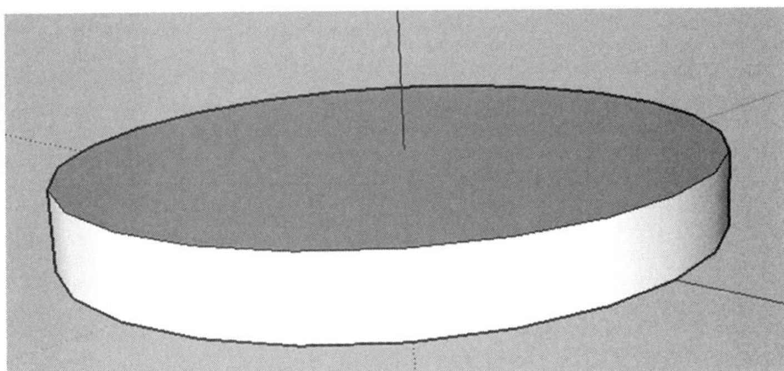

Clicking the Control key with the Push/Pull tool will push/pull multiple copies of the same face, which makes it convenient when a face of the same measurements needs to be extruded.

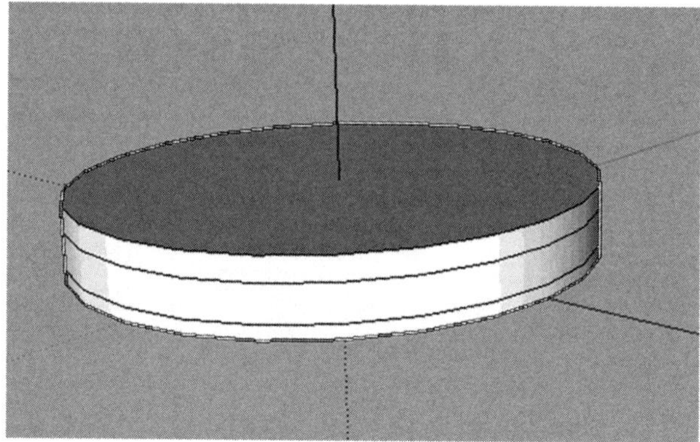

Geo-Location

SketchUp started as a Google product and, because of this, has benefits that no other 3D modeling program has. SketchUp is integrated with Google Earth, so users can add geographical location data to their model and then place the model into Google Earth to be viewed. Users can save their model as a Google Earth file that can be shared with other Google Earth users.

To add geographical data to a model created in SketchUp, open the SketchUp model. The model doesn't need to be completed—or even started. Models can be geolocated at any time in the design process. Next, click on File ➔ Geo-Location ➔ Add Location. This will open up the Google Earth window.

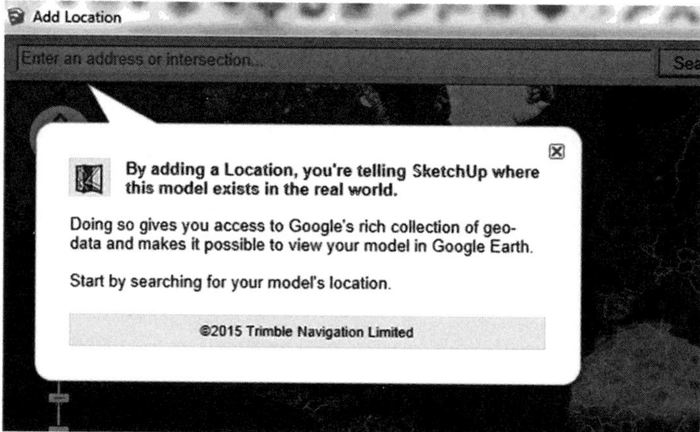

Now, find where the model will be located. This can easily be done by typing the address into the search box at the top of the window (an Internet connection is needed for this). Click on the Select Region box at the top left of the window.

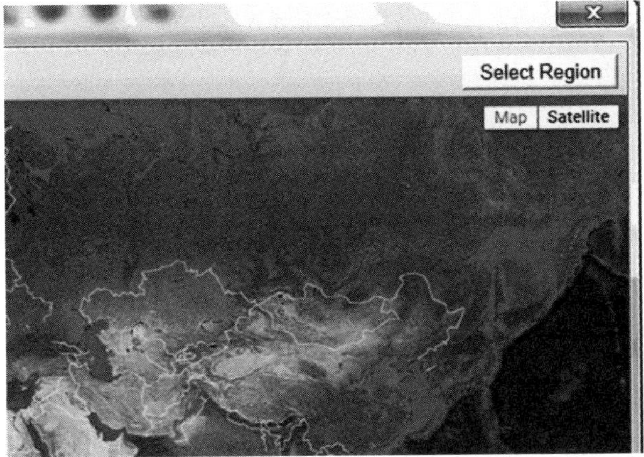

Click and drag the blue pins on the corners of the window to select the exact area for the geo-location. Click on the Grab button in the top left corner of the window to add a geo-location to the model.

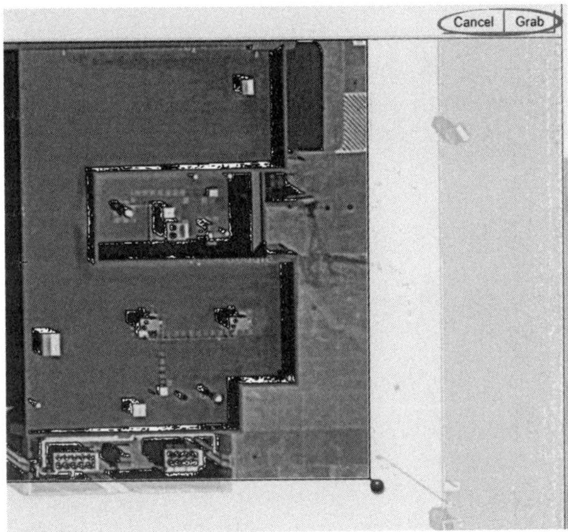

If the model is complete, users can utilize the Move and Rotate tools to move the model into vertical position on the terrain. Click File ➔ Geo-Location ➔ Show Terrain to view the model in the terrain. The most important thing to do at this point is to make sure the model is sitting on the terrain, not floating above it. Click File ➔ Save to save a copy of the model in the terrain.

3D Warehouse

3D Warehouse is a partner website to SketchUp and it hosts a catalog of millions of SketchUp models created by users from around the world. Users can upload completed models to 3D Warehouse and share them with other users, or make them private. Using 3D Warehouse as a tool to save and organize models, even if no one else can see them, can be useful for users of all skill levels. Browsing through the models is a great way for beginning modelers to begin to understand the possibilities of modeling and how models are put together. To access the 3D Warehouse, go to 3DWarehouse.SketchUp.com, or, while in SketchUp, go to File ➔3D Warehouse ➔ Get Models.

Models in 3D Warehouse can be searched by entering a search term in the search box at the top of the window. Models in 3D Warehouse that match the search terms will appear. Clicking on a picture of a model will open a larger picture of the model, along with more information.

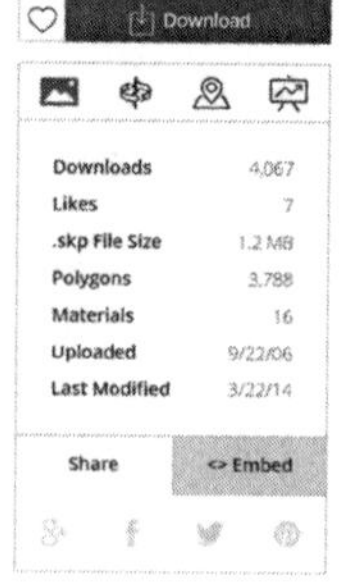

Information provided include how many times the model has been downloaded by users, the date it was uploaded by the original designer, and the dates of any modifications that have been made by the original designer. This window also shows the number of polygons (or faces) of the model. The greater the number of polygons in a model, the longer it will take to upload the model into SketchUp. After clicking the red Download button, users can choose whether to download the model into SketchUp or as a .KMZ file that can be placed in Google Earth.

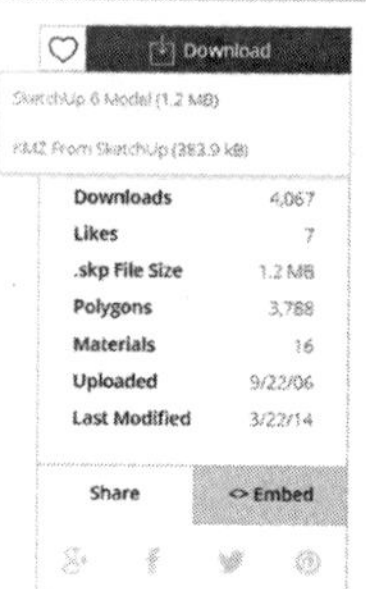

Clicking on the SketchUp file will open the model directly in the SketchUp program. Models can be virtually taken apart by students to get a better understanding of how models can be built.

Uploading Models into SketchUp to Share with Others

To save a completed model in the 3D Warehouse to share with others, open the model to be shared and click on File ➔ 3D Warehouse ➔ Share Model. The login screen for 3D Warehouse will appear. Users will need to create a Google account to upload a model into 3D Warehouse if they don't already have one. Most school districts now provide e-mail addresses for students, and these can be used for 3D Warehouse. Enter the Google account information and click on the Sign In button. This will open the Upload to 3D Warehouse window.

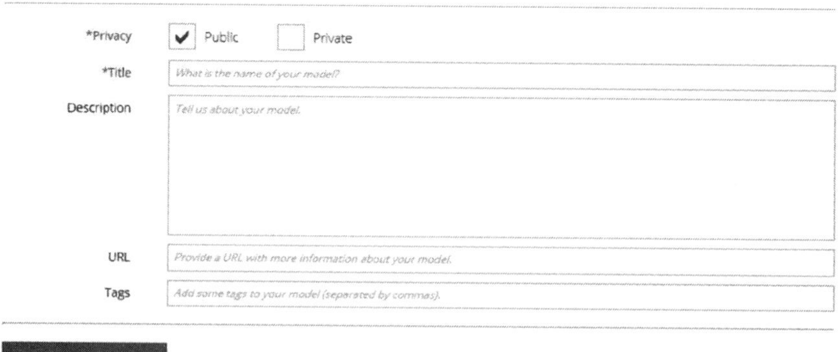

Privacy – Users must choose whether to make their model public or private. Choosing to make the model private will ensure the model is only seen by the user and cannot be downloaded by other users. Choose public will allow the model to be downloaded, modified, and 3D printed by any other 3D Warehouse user.

Title – Entering a descriptive title for the model will help other users search for it or, if the model is private, help the user organize and easily locate the model again.

Description – The description will be shown when the model is clicked on by those searching for models. A full and complete description of the model will help other users during their search for models to upload.

URL – A URL to a designer's website or to a site that contains more information about the model can be entered.

Tags – Tags are simply words that describe the model. Tags enable users to search more efficiently. When it comes to tags, the more tags the better. A comprehensive list of tags will ensure more people will be able to find the model. A great way to understand tags is by looking at tags that others have created for their models.

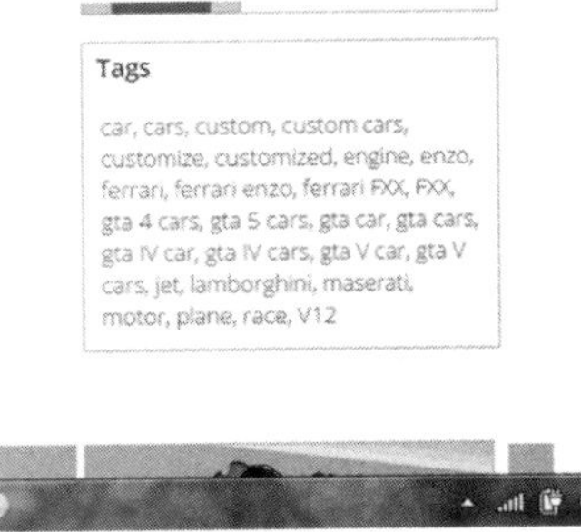

To finish uploading the model into 3D Warehouse, click on the red Upload button. After the model is uploaded, a picture of the model will appear, along with the information entered about the model. It usually takes less than 5 minutes for the model to appear in 3D Warehouse. After this short time, the model will be included in searches and can be uploaded by other users if the model is a public one.

Chapter 7 Key Points

SketchUp can be downloaded at sketchup.com.

There are two versions: SketchUp Pro, a paid version, and SketchUp Make, a free version that has more than enough modeling capabilities for students.

SketchUp also offers free Educator licenses. To receive an Educator license for SketchUp Pro for free, the educator must teach at an accredited educational institution and be able to provide a current faculty identification card.

The workspace includes red, blue, and green lines that represent the *x*, *y*, and *z* modeling axes. These lines help the user gain a sense of direction during the design process. These colored axis lines do not show up when the design is exported.

The Getting Started toolbar contains most of the tools needed to design a variety of 3D models.

The Status Bar shows whether the model is geo-located, shows credits for designs if the model has been imported from 3D Warehouse, and provides quick access to the Instructor, which is the SketchUp help menu.

3D Warehouse is a partner website to SketchUp and it hosts a catalog of millions of SketchUp models created by users from around the world. Users can upload completed models to 3D Warehouse and share them with other users, or make them private.

To save a completed model in the 3D Warehouse to share with others, open the model to be shared and click on File → 3D Warehouse → Share Model.

8

3D Modeling Using iPads and Android Tablets

A variety of 3D modeling apps can be used to create 3D models on an iPad or other tablet and then be sent to a 3D printer for printing. Many of the 3D modeling apps allow users to find models that have already been designed by other users and then send them to a 3D printer straight from the tablet. Fewer apps allow users to conduct actual creation of 3D models on the tablet. This chapter will cover a few 3D modeling apps in depth and provide step-by-step guidance on using these apps to design and print 3D objects.

Apps

123D Sculpt™

Available on iTunes for Apple devices

123D Sculpt is an app that students of all ages enjoy using to design objects that are only limited by the student's imagination.

When the 123D Sculpt app is opened, users have the choice to sculpt a creature, geometry, or object. Tapping on a category at the bottom of the screen opens sculpting options within that category. To begin sculpting, tap on one of the options and it will open in a design window.

There are many sculpting options for users on the design toolbar. Tapping on an icon in the toolbar prompts a description of what tapping on that icon does. Sculpting brush options include adding by pulling out, adding by layering up, removing by pushing in, smoothing areas out, sharpening soft edges, flattening areas, inflating areas, and dragging on areas to move and deform.

Tapping the paint brush icon on the toolbar prompts a color square to appear in the bottom right corner of the screen. Tapping this color square opens a vast menu of paint options. Users can choose a paint color and what type of brush to use. This screen also includes the Brush Size and Opacity bars. To paint the selected object, the user rubs their finger along the area to be painted.

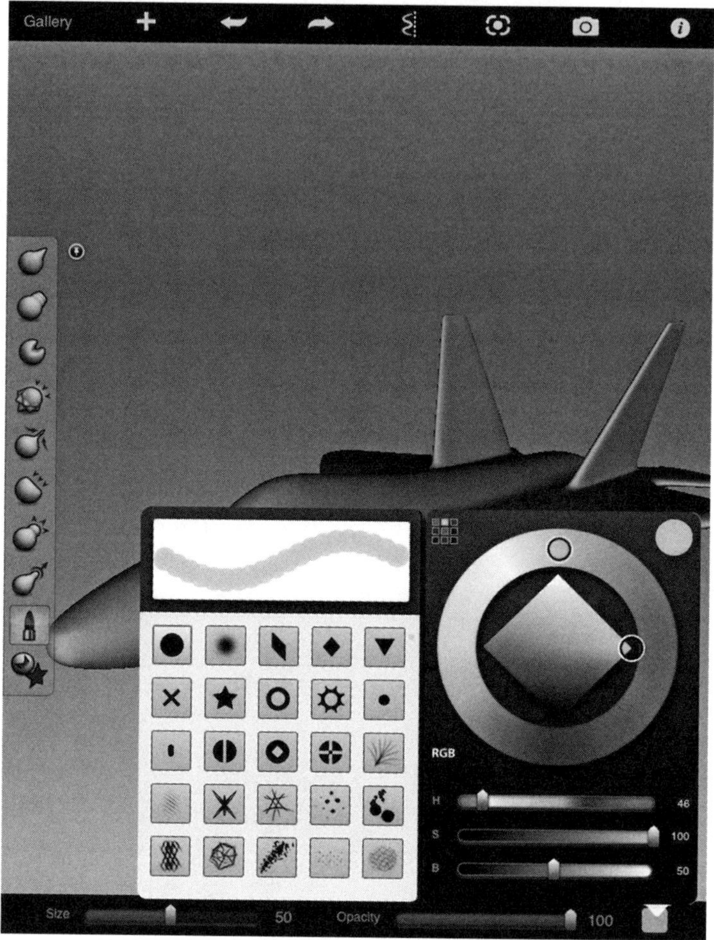

After painting, the object can be added to the Gallery. To do this, tap on the Gallery button in the top left corner. Users then have the option to Save the design. Saving the design will automatically add it to the Gallery. The design can then be shared by choosing the Save to Community option.

Autodesk screen shots reprinted with the permission of Autodesk, Inc.

Blokify – 3D Printing and Modeling
Available on iTunes for Apple devices

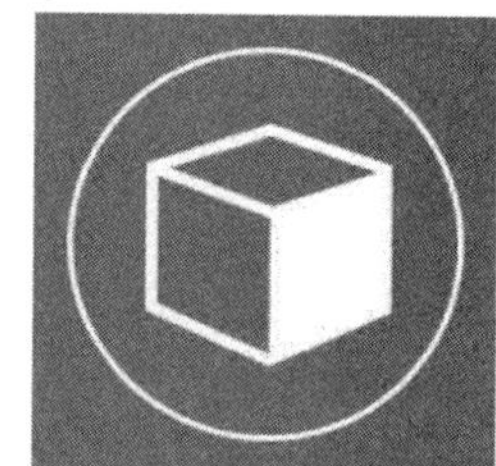

The Blokify app is free on iTunes. Blokify allows users to build objects using bloks with or without guided prompts. The main menu is accessed by tapping the arrow button in the top left corner of the screen. The menu has buttons for beginning a new structure, sharing, 3D printing, or changing the environment.

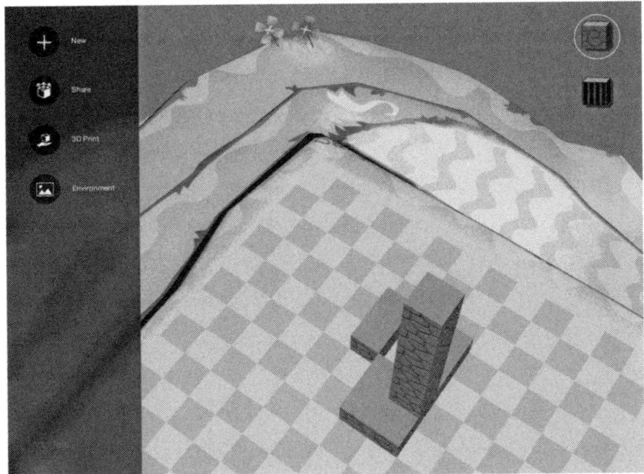

Blokify has two modes: challenge and free play. Upon opening for the first time, the app prompts the user to begin creating a model using the Challenge mode. In challenge mode, users place real bloks in place of blue ghost bloks.

The real challenge begins when the ghost bloks start disappearing quickly and the user has to really pay attention to where the bloks were before they disappeared. Blokify provides guidance along the way so students can learn it easily.

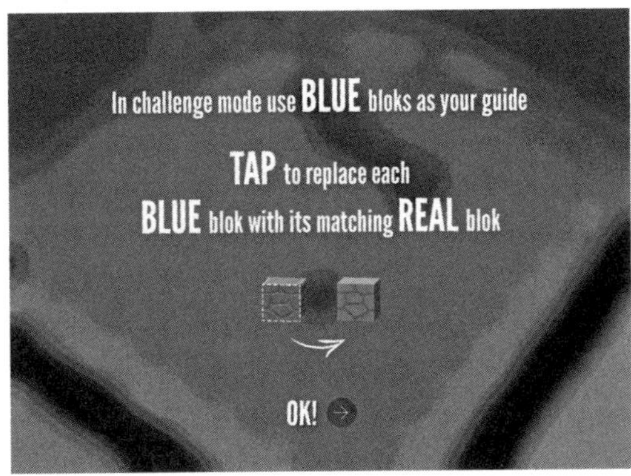

Although the app is free, there are some in-app purchases that can be added to add variety to building structures and environment. After a structure is built, the models can be ordered online or e-mailed. If the .STL file is e-mailed, the file can be saved, opened, and printed using the 3D printing software for the particular 3D printer being used. Models can be shared with others by e-mailing a link, which can be opened and saved by the recipient.

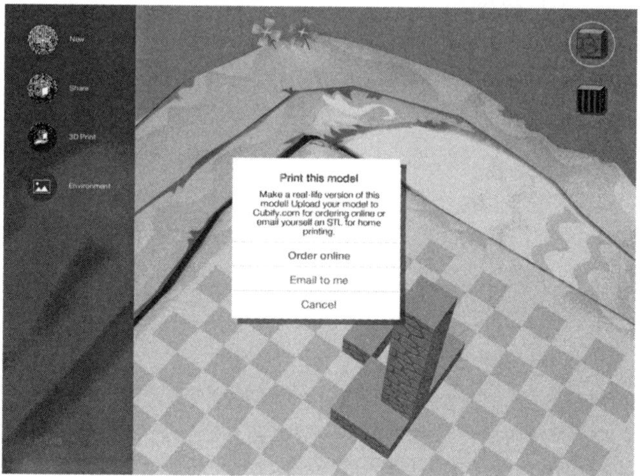

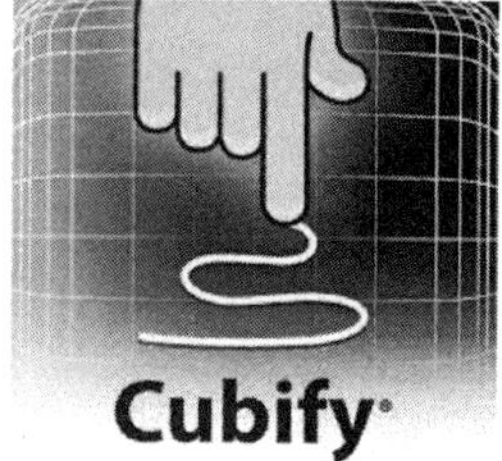

Cubify Draw

Available on iTunes for Apple devices

Cubify Draw is a free app that is available on the iTunes store. Cubify Draw was created by 3D Systems, the company that also makes the Cube, ProJet series, and Ekocycle 3D printers.

The app opens to a drawing screen where users can use a finger or stylus to draw an object.

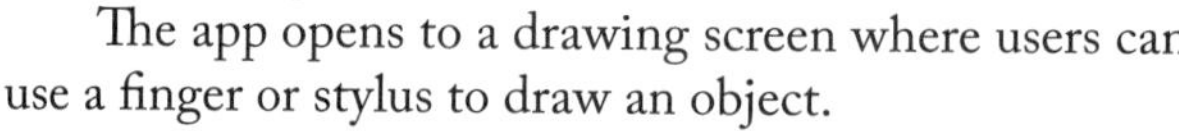

The menu at the bottom of the screen includes buttons to:

clear the screen

change the drawing line between thick and thin

connect the two ends of a line, and

upload an image.

Tapping the Choose Image button allows users to upload images to the screen and use their finger to draw around the image.

After drawing a design, press the Make it 3D button at the far right of the bottom menu bar. This will turn the drawing into a 3D model. The model can be adjusted for height, fill, and thickness.

Tapping the 3D Print button will prompt users to Upload or Email the design.

Press the Upload button to send the design to Cubify.com and users can choose to have 3D Systems print the object for them. Choosing the Email option will allow users to e-mail the .STL file to their own e-mail address, save it to the computer desktop, and then open and upload it with the 3D printing software for the particular printer being used.

Pressing the "i" button at the top right of the screen opens the help screen, which explains all buttons in the app.

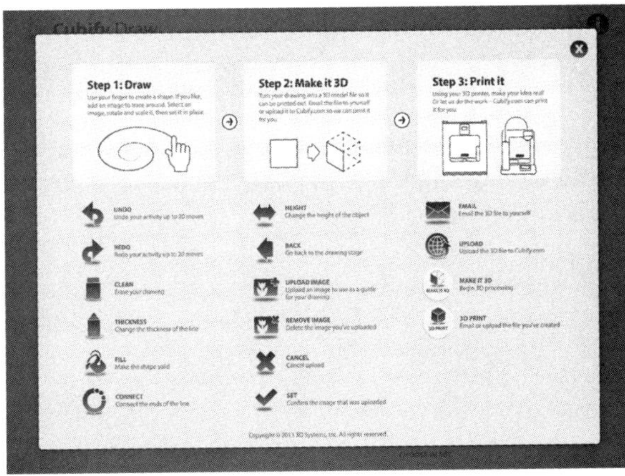

Mecube

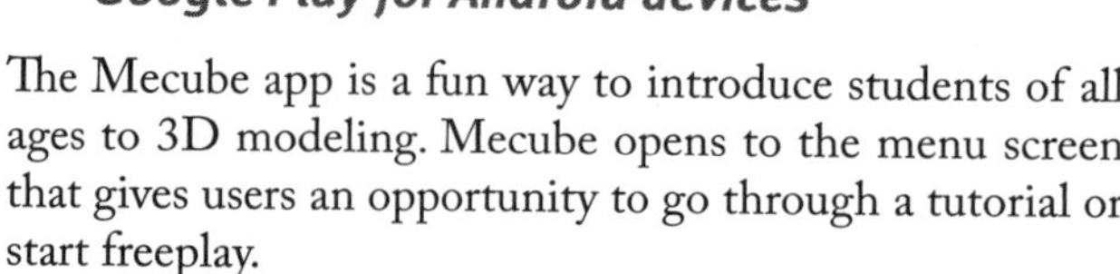

Available on iTunes for Apple devices and on Google Play for Android devices

The Mecube app is a fun way to introduce students of all ages to 3D modeling. Mecube opens to the menu screen that gives users an opportunity to go through a tutorial or start freeplay.

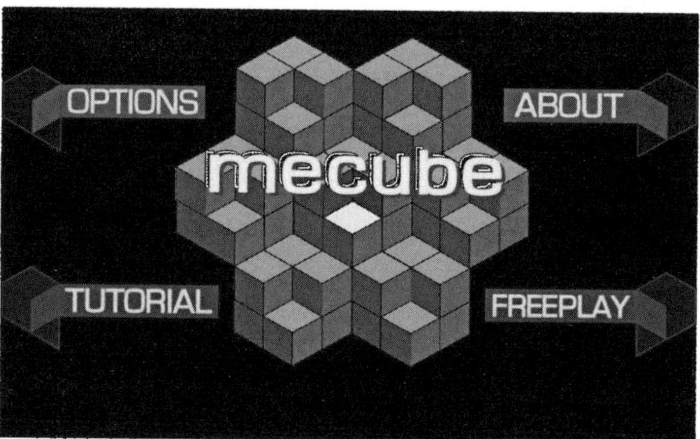

Every Mecube design begins with the basic cube design. Manipulating the cube will rotate it for different views.

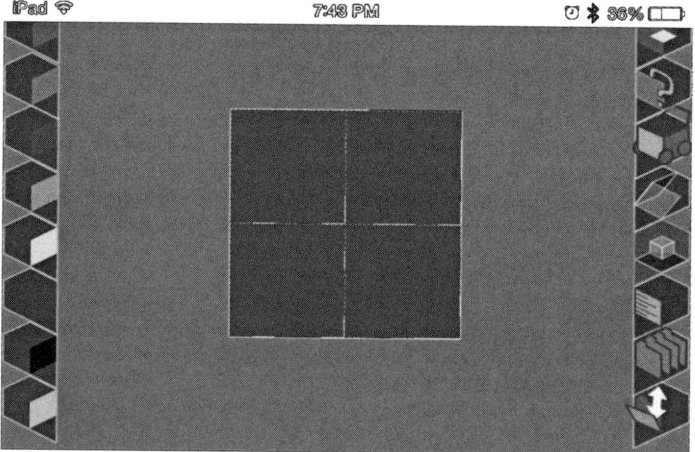

Color can be added to the cube by tapping a color tab on the left side of the screen and then tapping a block in the cube. For more saturated color, tap the block multiple times.

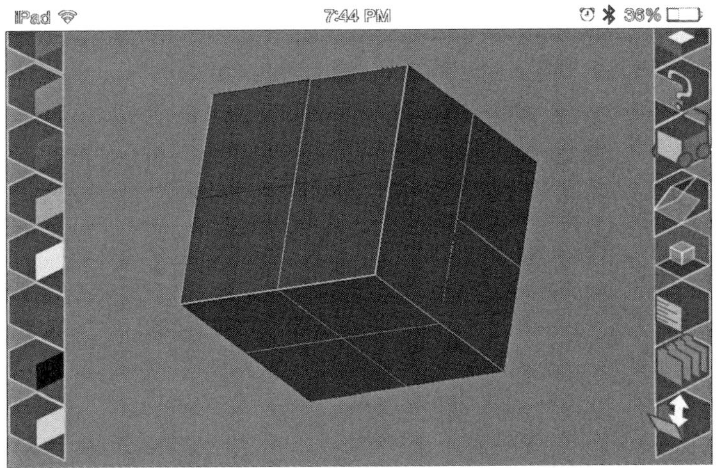

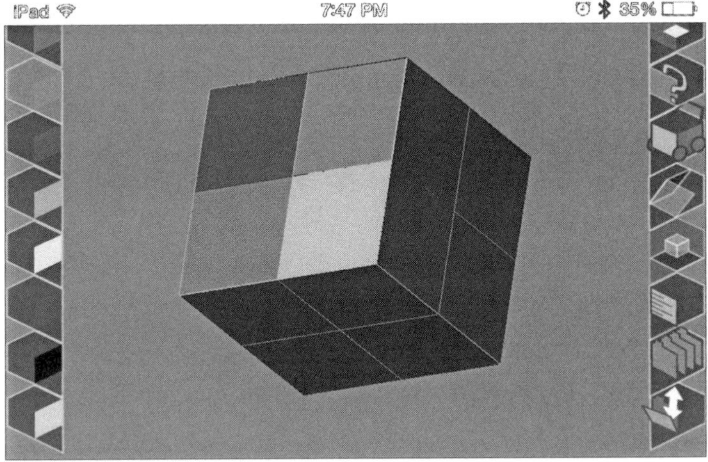

The menu on the right side of the screen allows users to scale the model, add or delete blocks, and set model parameters.

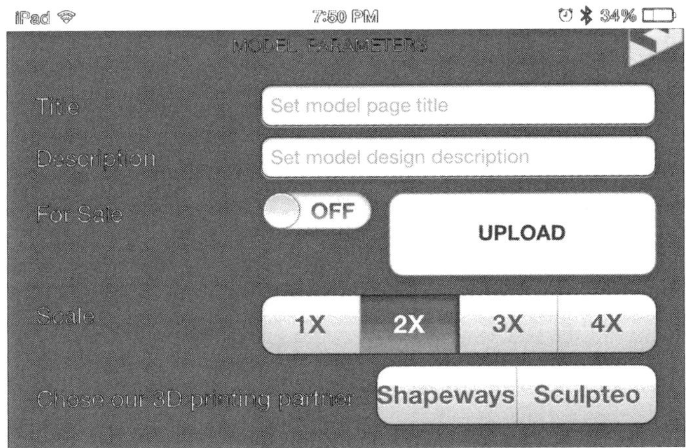

Selecting a 3D printing partner and tapping the Upload button will send the model to that printing partner. Users can then choose the material they want the model printed in and have the model printed by Shapeways or Sculpteo for a fee. Usually this option would not be used in a school setting due to the cost of printing the designs, but the modeling experience alone makes this app worthwhile.

Monstermatic

Available on iTunes for Apple devices

Monstermatic is a relatively new app that incorporates gaming with 3D modeling. Users create their own monsters that have swappable parts. Monstermatic is an example of a program that got its start on Kickstarter, the crowdsourcing website—247 backers donated over $30,000 to support the release of Monstermatic on iTunes.

To begin with Monstermatic, open the app and tap on the Design a Monster Here button. This will initiate a tutorial that will take users through the entire fun design process.

Users can customize the eyes, mouth, hands, feet, clothing, and accessories and can utilize the paint feature to change the color of all or part of their monster by scrolling through the design bar at the bottom of the screen. Tapping on the storefront icon allows users to order a 3D print figurine of the monster or order a t-shirt with the monster on it. As a part of the game of creating a monster, users can earn coins to help cover the cost of the items.

Chapter 8 Key Points

3D modeling apps can be used to create 3D models on an Apple or Android tablet and then be sent to a 3D printer for printing.

Many 3D modeling apps allow users to find models that have already been designed by others and can be sent to a 3D printer straight from the tablet.

Some apps allow users to actually create 3D models on the tablet.

Most 3D modeling apps are free but some offer in-app purchases.

9

3D Modeling and Printing in the Math Curriculum

The following chapters contain lesson ideas that librarians can utilize to integrate 3D design and printing to effectively teach 21st-century skills. These lesson ideas should be used as a starting point with students. Begin with 3D design and printing in the library by selecting one of the lesson ideas that is appropriate for your students. Seek out those teachers who are willing to collaborate. Often, collaborating with just one teacher or grade level will show others in your building how wonderful 3D modeling and printing is for students. Teachers may be hesitant at first, but when they see the benefits for students and how much students enjoy creating and printing their own models, they will soon jump at the chance to collaborate on a 3D printing project.

It will take most students a few times to get the hang of 3D design, so don't get frustrated if it seems to move slowly at first. After students get the hang of it, design will move much faster. Using design software that students can access at home will help students learn it faster. If using a web-based design program, such as TinkerCad or 3DTin, students will be able to work on designs at home if they have access to the Internet. Giving students access to the design software so they can practice is key to learning 3D modeling. Librarians can make it easier for students by posting links to modeling programs on the library website. Librarians can also create bookmarks for students (or have students design and create the bookmarks) that list the URLs for the programs and give them out freely.

When students come to the library for the first time for 3D design, allow plenty of time to show students key aspects of the program. Often, showing students how to get started with the program, then showing basic key elements of how to use the program is all students need. Encourage students to help other students when they can. Often students will feel more comfortable asking peers for help, so allowing for some conversation during the modeling and design process, as long as the students are staying on task, can be helpful.

⁕ ⁕ ⁕

The following lesson ideas are examples of how 3D modeling and printing can be utilized as a curriculum tool. The lessons are associated with grade levels in order to establish the connection with national standards. Librarians should feel free to use these lessons as springboards for their own ideas or to tweak the lessons and use them with another grade level than the one listed. For example, a lesson idea presented for third grade may inspire another lesson idea for a high school class. The following lessons are in no way meant to be all-inclusive, as the variety of ways 3D modeling and printing can be used in the curriculum is endless.

3D Shapes

Topic: Counting and Cardinality, Understanding Addition and Understanding Subtraction

Grade Level: Kindergarten

Related Standards:

AASL Standards for the 21st-Century Learner

1.1.2 - Use prior and background knowledge as context for new learning.

2.1.6 - Use the writing process, media and visual literacy, and technology skills to create products that express new understandings.

2.1.3 - Use strategies to draw conclusions from information and apply knowledge to curricular areas, real-world situations, and further investigations.

Common Core Standards

CCSS.MATH.CONTENT.K.CC.A.2 - Count forward beginning from a given number within the known sequence (instead of having to begin at 1).

CCSS.MATH.CONTENT.K.OA.A.1 - Represent addition and subtraction with objects, fingers, mental images, drawings, sounds (e.g., claps), acting out situations, verbal explanations, expressions, or equations.

CCSS.MATH.CONTENT.K.OA.A.2 - Solve addition and subtraction word problems, and add and subtract within 10, e.g., by using objects or drawings to represent the problem.

How to Integrate 3D Modeling and Printing

3D modeling allows students to personalize the counting experience by allowing them to design individualized 3D shapes and then print them. In this activity, each student will design one shape, and then each shape can be printed so there is a set for students to use in the classroom or library. Students will love using the shapes they have created and personalized for counting practice. Students can also use the objects as manipulatives when solving word problems. To make the selection process go faster, the librarian can write the shapes on index cards or pieces of paper or have students draw the shapes on pieces of paper. Students can draw a card from the set to see which shape they will be responsible for designing. Having a class set of manipulatives unique to that class allows students to take pride in creating something the whole class can use. These shapes will be printed in the library and shared with the classroom teacher for students to enjoy using in the classroom.

3D Shapes with Defining Attributes

Topic: Geometry

Grade Level: 1

Related Standards:

AASL Standards for the 21st-Century Learner

1.1.2 - Use prior and background knowledge as context for new learning.

2.1.3 - Use strategies to draw conclusions from information and apply knowledge to curricular areas, real-world situations, and further investigations.

Common Core Standard

CCSS.MATH.CONTENT.1.G.A.1 - Distinguish between defining attributes (e.g., triangles are closed and three-sided) versus non-defining attributes (e.g., color, orientation, overall size); build and draw shapes to possess defining attributes.

How to Integrate 3D Modeling and Printing

After discussing the difference between defining attributes and nondefining attributes, students will utilize 3D modeling software to design shapes exhibiting defining attributes. Students will work in small groups, and the librarian will assign the shape each group will design to ensure variety. After printing the shapes, students will present their objects to peers, describing the defining characteristics of the object. Objects can be displayed in the library or shared with the classroom teacher for classroom use.

Tangrams

Topic: Geometry

Grade Level: 1

Related Standards:

AASL Standards for the 21st-Century Learner

1.1.2 - Use prior and background knowledge as context for new learning.

2.1.2 - Organize knowledge so that it is useful.

2.1.3 - Use strategies to draw conclusions from information and apply knowledge to curricular areas, real-world situations, and further investigations.

2.1.6 - Use the writing process, media and visual literacy, and technology skills to create products that express new understandings.

Common Core Standard

CCSS.MATH.CONTENT.1.G.A.2 - Compose two-dimensional shapes (rectangles, squares, trapezoids, triangles, half-circles, and quarter-circles) or three-dimensional shapes (cubes, right rectangular prisms, right circular cones, and right circular cylinders) to create a composite shape, and compose new shapes from the composite shape.

How to Integrate 3D Modeling and Printing

Before the lesson, the librarian will find a tangram model for students to print, such as the complete tangram set by sgosiaco on Thingiverse.com at www.thingiverse.com/thing:184076.

Incorporate literature and integrate multiple literacies by reading *Grandfather Tang's Story* by Ann Tompert. Students will print a set of tangrams prior to reading, either individually or in small groups. The librarian will read the book one time through and then re-read slowly, allowing time for students to recreate the tangram animals in the story. The school librarian can display the pages on a projector for all students to see at one time and will probably need to allow for plenty of time for students to work, especially with the more complex animals. The librarian can point out or show the shapes to students if more help is needed. After students complete an animal from the book, allow students to use the tangram sets to design their own animals. Some of the tangram sets can be shared with the classroom teacher but a few of the sets should be kept in the library along with an extra copy of *Grandfather Tang's Story*, because this will become a favorite activity of students in all grade levels when they visit the library.

Halves and Quarters

Topic: Geometry

Grade Level: 1

Related Standards:

AASL Standards for the 21st-Century Learner

1.1.2 - Use prior and background knowledge as context for new learning.

2.1.1 - Continue an inquiry-based research process by applying critical-thinking skills (analysis, synthesis, evaluation, organization) to information and knowledge in order to construct new understandings, draw conclusions, and create new knowledge.

2.1.2 - Organize knowledge so that it is useful.

2.1.3 - Use strategies to draw conclusions from information and apply knowledge to curricular areas, real-world situations, and further investigations.

2.1.6 - the writing process, media and visual literacy, and technology skills to create products that express new understandings.

Common Core Standard

CCSS.MATH.CONTENT.1.G.A.3 - Partition circles and rectangles into two and four equal shares, describe the shares using the words *halves*, *fourths*, and *quarters*, and use the phrases *half of*, *fourth of*, and *quarter of*. Describe the whole as two of, or four of the shares. Understand for these examples that decomposing into more equal shares creates smaller shares.

How to Integrate 3D Modeling and Printing

Using 3D modeling software, students will design circles and rectangles, either individually or in small groups. Direct students to divide the shape in half and then into quarters using the cut tool in the modeling software. Students can repeat until the librarian is satisfied they understand the concept of dividing objects into two and four equal shares. Shapes designed by students can be 3D printed so they can manipulate their creations by hand.

Measurement

Topic: Measure and estimate lengths in standard units.

Grade Level: 2

Related Standards:

AASL Standards for the 21st-Century Learner

1.1.2 - Use prior and background knowledge as context for new learning.

1.1.8 - Demonstrate mastery of technology tools for accessing information and pursuing inquiry.

2.1.2 - Organize knowledge so that it is useful.

3.1.4 - Use technology and other information tools to organize and display knowledge and understanding in ways that others can view, use, and assess.

Common Core Standards

CCSS.MATH.CONTENT.2.MD.A.1 - Measure the length of an object by selecting and using appropriate tools such as rulers, yardsticks, meter sticks, and measuring tapes.

CCSS.MATH.CONTENT.2.MD.A.2 - Measure the length of an object twice, using length units of different lengths for the two measurements; describe how the two measurements relate to the size of the unit chosen.

CCSS.MATH.CONTENT.2.MD.A.3 - Estimate lengths using units of inches, feet, centimeters, and meters.

How to Integrate 3D Modeling and Printing

Students will design objects and use the measurement tool in the software program to create an object of a size specified by the librarian. Students record the measurements for length, width, and height of their objects. After each object is 3D printed, students will estimate the length in inches of objects printed by each student and then measure the object with a ruler to see which estimate was closest. Many 3D modeling software programs show measurements of designed objects in Standard or Metric measurement. Each modeling software program is different, but most provide a way to toggle between millimeters and inches for measuring. For example, in Tinkercad, users click on Edit Grid in the lower left corner and select the unit they wish to measure in.

Fractions

Topic: Numbers and Operations

Grade Level: 3

Related Standards:

AASL Standards for the 21st-Century Learner

2.1.3 - Use strategies to draw conclusions from information and apply knowledge to curricular areas, real-world situations, and further investigations.

2.1.6 - Use the writing process, media and visual literacy, and technology skills to create products that express new understandings.

Common Core Standards

CCSS.MATH.CONTENT.3.NF.A.3 - Explain equivalence of fractions in special cases, and compare fractions by reasoning about their size.

CCSS.MATH.CONTENT.3.NF.A.3.A - Understand two fractions as equivalent (equal) if they are the same size, or the same point on a number line.

CCSS.MATH.CONTENT.3.NF.A.3.B - Recognize and generate simple equivalent fractions, e.g., 1/2 = 2/4, 4/6 = 2/3. Explain why the fractions are equivalent, e.g., by using a visual fraction model.

How to Integrate 3D Modeling and Printing

Students will search for a model of a fraction block on a site such as Thingiverse.com or the librarian can find this before the lesson. Students will print fraction blocks either individually or in small groups and then manipulate the printed fraction blocks to compare fractions. To save time, the librarian can print a fraction block before the lesson and then print one while students are in the library. One of the fraction blocks can be shared with the classroom teacher and the other one kept in the library for future use.

Parallel and Perpendicular Lines

Topic: Draw and identify lines and angles, and classify shapes by properties of their lines and angles.

Grade Level: 4

Related Standards:

AASL Standards for the 21st-Century Learner

2.1.6 - Use the writing process, media and visual literacy, and technology skills to create products that express new understandings.

Common Core Standard

CCSS.MATH.CONTENT.4.G.A.2 - Classify two-dimensional figures based on the presence or absence of parallel or perpendicular lines, or the presence or absence of angles of a specified size. Recognize right triangles as a category, and identify right triangles.

How to Integrate 3D Modeling and Printing

After reviewing characteristics of parallel and perpendicular lines, students will design shapes and count the parallel and perpendicular sets of lines and the number of right angles in the object. Next, students will add angles of specified degrees to their design. For example, have students add a 45-degree angle to their object or change the measurement of an existing angle to 45 degrees. Follow with angles of other measurements. Objects can be printed or saved in the modeling software.

Volume

Topic: Geometric measurement

Grade Level: 5

Related Standards:

AASL Standards for the 21st-Century Learner

1.1.1 - Follow an inquiry-based process in seeking knowledge in curricular subjects, and make the real-world connection for using this process in own life.

2.1.3 - Use strategies to draw conclusions from information and apply knowledge to curricular areas, real-world situations, and further investigations.

Common Core Standards

CCSS.MATH.CONTENT.5.MD.C.3 - Recognize volume as an attribute of solid figures and understand concepts of volume measurement.

CCSS.MATH.CONTENT.5.MD.C.4 - Measure volumes by counting unit cubes, using cubic cm, cubic in, cubic ft, and improvised units.

CCSS.MATH.CONTENT.5.MD.C.5 - Relate volume to the operations of multiplication and addition and solve real world and mathematical problems involving volume.

CCSS.MATH.CONTENT.5.MD.C.5.B - Apply the formulas $V = l \times w \times h$ and $V = b \times h$ for rectangular prisms to find volumes of right rectangular prisms with whole-number edge lengths in the context of solving real world and mathematical problems.

How to Integrate 3D Modeling and Printing

Review the concept of volume of solid figures. Students will design rectangular prisms in the 3D modeling software and then find the volume of the object using the formulas $V = l \times w \times h$ and $V = b \times h$. The librarian can integrate problem-based learning by providing students with a real-world scenario involving rectangular prisms, such as a scenario in which students are engineers designing a box to contain a particular object. To complete this project, students will work in small groups to design a container that would hold an object, such as a book of a certain size. Because the design wouldn't need to be printed, the model will be saved in the modeling software program.

Ratio

Topic: Understand ratio concepts and use ratio reasoning to solve problems.

Grade Level: 6

Related Standards:

AASL Standards for the 21st-Century Learner

2.3.1 - Connect understanding to the real world.

Common Core Standards

CCCS.MATH.CONTENT.6.RP.3.A - Use ratio and rate reasoning to solve real-world and mathematical problems, e.g., by reasoning about tables of equivalent ratios, tape diagrams, double number line diagrams, or equations

CCSS.MATH.CONTENT.6.RP.A.3.D - Use ratio reasoning to convert measurement units; manipulate and transform units appropriately when multiplying or dividing quantities.

How to Integrate 3D Modeling and Printing

Review the concept of ratios and how to convert units by multiplying and dividing quantities. Before the lesson, the librarian will design a simple object using 3D modeling software and save it in the modeling program. In small groups, students will open the file and design a new object by scaling the original object up or down in size as designated by the librarian. The designs can be saved in the modeling program or the librarian can select one or two objects to be printed.

Geometrical Figures

Topic: Geometry—Draw and describe geometrical figures and describe the relationships between them.

Grade Level: 7

Related Standards:

AASL Standards for the 21st-Century Learner

1.1.2 - Use prior and background knowledge as context for new learning.

Common Core Standards

CCSS.MATH.CONTENT.7.G.A.1 - Solve problems involving scale drawings of geometric figures, including computing actual lengths and areas from a scale drawing and reproducing a scale drawing at a different scale.

CCSS.MATH.CONTENT.7.G.A.2 - Draw (freehand, with ruler and protractor, and with technology) geometric shapes with given conditions. Focus on constructing triangles from three measures of angles or sides, noticing when the conditions determine a unique triangle, more than one triangle, or no triangle.

CCSS.MATH.CONTENT.7.G.A.3 - Describe the two-dimensional figures that result from slicing three-dimensional figures, as in plane sections of right rectangular prisms and right rectangular pyramids.

How to Integrate 3D Modeling and Printing

In small groups, students will measure the library, playground, courtyard, or any other space or object around the school and design and scale the measurements on graph paper. Next, students will design the scaled version of the measured space or object using 3D modeling software. Students could design multiple versions of the measured space or object on 3D modeling software using different scales.

Students will design geometric shapes with specific measures of angles and sides using 3D modeling software. The librarian will give students measures of three angles and sides and students will design that triangle. Students will decide what type of triangle the shape is by evaluating the angle measures.

Volume of Cylinders, Cones, and Spheres

Topic: Solve real-world and mathematical problems involving volume of cylinders, cones, and spheres.

Grade Level: 8

Related Standards:

AASL Standards for the 21st-Century Learner

1.1.2 - Use prior and background knowledge as context for new learning.

Common Core Standard

CCSS.MATH.CONTENT.8.G.C.9 - Know the formulas for the volumes of cones, cylinders, and spheres and use them to solve real-world and mathematical problems.

How to Integrate 3D Modeling and Printing

Students, working individually or in small groups, will design cones, cylinders, and spheres using 3D modeling software. The librarian can give students measurements for the shapes or the size can be student choice. If student choice, make sure students understand the maximum print size of the printer being used. Students will 3D print the objects and then use volume formulas to find the volume of each shape designed by students.

Pythagorean Theorem

Topic: Geometry

Grade Level: 8

Related Standards:

AASL Standards for the 21st-Century Learner

1.1.2 - Use prior and background knowledge as context for new learning.

Common Core Standards

CCSS.MATH.CONTENT.8.G.B.7 - Apply the Pythagorean Theorem to determine unknown side lengths in right triangles in real-world and mathematical problems in two and three dimensions.

CCSS.MATH.CONTENT.8.G.B.8: - Apply the Pythagorean Theorem to find the distance between two points in a coordinate system.

How to Integrate 3D Modeling and Printing

Present students with a scenario in which they have to design an object such as a television or computer monitor with a specific height and width using 3D modeling software. Students will then calculate the diagonal length of the television or monitor screen of the object they have designed.

Students will tour the school to determine if an access ramp needs to be added to any location around the school that has an entry above ground level. If time is limited, the librarian can give students this scenario instead of sending students around the school. Students will research information on access ramps and design a ramp using 3D modeling software that is in compliance with the Americans with Disabilities Act (ADA) requirements. An extension of this activity is that students can measure the base, height, slope, and angle of existing ramps around the school to determine if they are ADA compliant.

Students will design a skate park with ramps. Ramps can be scaled smaller and 3D printed and students can then use the Pythagorean theorem to determine rise, run, and ramp length. Students can evaluate each ramp to determine if their scaled ramp would be an appropriate model for a real-life skate park.

Rotation, Reflection, and Translation

Topic: Understand congruence and similarity using physical models, transparencies, or geometry software.

Grade Level: 8

Related Standards:

AASL Standards for the 21st-Century Learner

1.1.2 - Use prior and background knowledge as context for new learning.

Common Core Standards

CCSS.MATH.CONTENT.8.G.A.1 - Verify experimentally the properties of rotations, reflections, and translations

CCSS.MATH.CONTENT.HSG-CO.A.3 - Given a rectangle, parallelogram, trapezoid, or regular polygon, describe the rotations and reflections that carry it onto itself.

CCSS.MATH.CONTENT.HSG-CO.A.5 - Given a geometric figure and a rotation, reflection, or translation, draw the transformed figure using, e.g., graph paper, tracing paper, or geometry software. Specify a sequence of transformations that will carry a given figure onto another.

How to Integrate 3D Modeling and Printing

Students will design a basic shape using 3D modeling software. If time is limited, the librarian will design or find a model for students to use before the lesson. Students will rotate, reflect, and translate. Students will also describe sizes, positions, and orientations of shapes under transformations such as flips, turns, slides, and scaling.

Transformations

Topic: Experiment with transformations in the plane

Grade Levels: 9–12

Related Standards:

AASL **Standards for the 21st-Century Learner**

2.1.3 - Use strategies to draw conclusions from information and apply knowledge to curricular areas, real-world situations, and further investigations.

Common Core Standard

CCSS.MATH.CONTENT.HSG.CO.A.2 - Represent transformations in the plane using, e.g., transparencies and geometry software; describe transformations as functions that take points in the plane as inputs and give other points as outputs. Compare transformations that preserve distance and angle to those that do not (e.g., translation versus horizontal stretch).

How to Integrate 3D Modeling and Printing

Students will design basic shapes in a modeling software program. Students will practice moving the shapes around the coordinate plane in the program according to different types of transformations. Students can use coordinates to represent translations as functions. Allow students to practice comparing rigid transformations that preserve distance and angle with transformations that do not.

Transformation Sequences

Topic: Experiment with transformations in the plane

Grade Levels: 9–12

Related Standards:

AASL Standards for the 21st-Century Learner

2.1.3 - Use strategies to draw conclusions from information and apply knowledge to curricular areas, real-world situations, and further investigations.

Common Core Standard

CCSS.MATH.CONTENT.HSG.CO.A.5 - Given a geometric figure and a rotation, reflection, or translation, draw the transformed figure using, e.g., graph paper, tracing paper, or geometry software. Specify a sequence of transformations that will carry a given figure onto another.

How to Integrate 3D Modeling and Printing

Allow students to use modeling software to show how a transformation maps one shape onto another. As an extension, have students discuss why manipulating figures in this way would be useful in the real world. For this activity, objects are just manipulated in the modeling program and not printed.

Coordinate Geometry

Topic: Use coordinates to prove simple geometric theorems algebraically

Grade Levels: 9–12

Related Standard:

Common Core Standard

CCSS.MATH.CONTENT.HSG.GPE.B.7 - Use coordinates to compute perimeters of polygons and areas of triangles and rectangles, e.g., using the distance formula.

How to Integrate 3D Modeling and Printing

Students will use the coordinate grid in a modeling software program to design polygons, triangles, and rectangles and then find the perimeters of designed figures. For this activity, objects don't need to be printed and can be saved in the modeling program, if desired.

2D to 3D

Topic: Visualize relationships between two-dimensional and three-dimensional objects

Grade Levels: 9–12

Related Standard:

Common Core Standard

CCSS.MATH.CONTENT.HSG.GMD.B.4 - Identify the shapes of two-dimensional cross-sections of three-dimensional objects, and identify three-dimensional objects generated by rotations of two-dimensional objects.

How to Integrate 3D Modeling and Printing

To help students connect 2D and 3D objects, they will design cubes, pyramids, cones, spheres, and cylinders using a modeling software program. Students will slice each object in different ways to see cross sections of each figure and the 2D shapes formed by each cross-section. Allow students to discuss what types of shapes the cross-sections make.

Real-World Problem Solving

Topic: Apply geometric concepts in real-world situations

Grade Levels: 9–12

Related Standards:

AASL Standards for the 21st-Century Learner

2.1.3 - Use strategies to draw conclusions from information and apply knowledge to curricular areas, real-world situations, and further investigations.

3.3.4 - Create products that apply to authentic, real-world contexts.

Common Core Standard

CCSS.MATH.CONTENT.HSG.MG.A.3 - Apply geometric methods to solve design problems (e.g., designing an object or structure to satisfy physical constraints or minimize cost; working with typographic grid systems based on ratios).

How to Integrate 3D Modeling and Printing

This is an opportunity to incorporate project-based learning by presenting students with a real-world design problem they need to solve. Present the scenario to students by displaying the scenario for all students to refer to as they work. Students will design objects based on the presented scenario and the objects will be printed. Scenarios are fun for students because it they involve some type of role playing and are fun for the librarian to engage in, as well. Some scenarios that can be presented are as follows:

Scenario: The principal or local city council member has requested that students design a playground for a school or the community. Students, in small groups, are assigned a specific piece of equipment they are responsible for designing, such as a slide, swingset, climbing structure, etc. Students will research safety guidelines for their piece of equipment and design a scaled model using a 3D modeling program.

Scenario: A local gardening group would like to add a new pergola to the city garden. Students will be divided into small groups and design a scaled model of a pergola that meets the group's requirements.

Scenario: A major cell phone manufacturer has contacted the school and they need students to design a case for their newly designed cell phone. Students are given the dimensions of the cell phone and will design a case that students their own age would be interested in purchasing.

10

3D Modeling and Printing in the Science Curriculum

Using 3D modeling and printing in the science curriculum not only engages students with a variety of learning styles, it gives librarians an innovative way to connect with the curriculum. The following lessons can be used with a wide range of grade levels to integrate 3D modeling and printing. Next Generation Science Standards have been included when possible to show how 3D modeling and printing can be connected to the science curriculum.

Improving on Everyday Objects

Topic: Engineering Design

Grade Levels: K–2

Related Standards:

Next Generation Science Standards

K-2-ETS1-1. Ask questions, make observations, and gather information about a situation people want to change to define a simple problem that can be solved through the development of a new or improved object or tool.

K-2-ETS1-2. Develop a simple sketch, drawing, or physical model to illustrate how the shape of an object helps it function as needed to solve a given problem.

K-2-ETS1-3. Analyze data from tests of two objects designed to solve the same problem to compare the strengths and weaknesses of how each performs.

How to Integrate 3D Modeling and Printing

Discuss with students how the shapes of everyday objects help their functions (examples: spoon and comb). Have students brainstorm problems that could be solved if only an improved model of an object could be designed. Students can design their model on paper and then on 3D modeling software such as Tinkercad.com. Objects could be 3D printed and tested to determine strengths and weaknesses.

Designing Simple Machines

Topic: Physics

Grade Levels: 2–3

Related Standards:

AASL Standards for the 21st-Century Learner

1.1.8 - Demonstrate mastery of technology tools to access information and pursue inquiry.

3.1.4 - Use technology and other information tools to organize and display knowledge and understanding in ways that others can view, use, and assess.

Next Generation Science Standard

3-PS2-1. Plan and conduct an investigation to provide evidence of the effects of balanced and unbalanced forces on the motion of an object.

How to Integrate 3D Modeling and Printing

Integrating 3D modeling and printing into a unit on simple machines is an effective way to engage students and let them design and create actual models of simple machines. Students can design their own simple machines using 3D modeling software or find models already created of each type of simple machine. Review characteristics of each type of simple machine and then allow students to work individually or in groups to design each type.

As an extension, have students think of and design an invention that uses at least one type of simple machine. Students can design the invention on paper and then move to the 3D modeling software. Inventions can be 3D printed and students can present their inventions to each other and to parents.

Project-Based Design

Topic: Engineering Design

Grade Levels: 3–5

Related Standards:

Next Generation Science Standards

3-5-ETS1-1. Define a simple design problem reflecting a need or a want that includes specified criteria for success and constraints on materials, time, or cost.

3-5-ETS1-2. Generate and compare multiple possible solutions to a problem based on how well each is likely to meet the criteria and constraints of the problem.

3-5-ETS1-3. Plan and carry out fair tests in which variables are controlled and failure points are considered to identify aspects of a model or prototype that can be improved.

How to Integrate 3D Modeling and Printing

Ensure students are familiar with the terms constraints and criteria. Give students a scenario in which they must design an object using 3D modeling software with a set of criteria and constraints. Constraints could be related to the amount or types of materials used, the length of time they have to complete the design, or could be related to the cost of materials used. If needed, the librarian could introduce one constraint at a time until finally introducing students to all three constraints of the project. After printing using a 3D printer, the objects could be weighed and each group would be "charged" the cost of PLA plastic filament used to make the object.

Scenarios don't have to be elaborate in order to engage students. A favorite of students is designing a carton or container that would keep an egg from breaking during shipping. After 3D printing the objects, students can perform tests to determine how designs could be improved and present these improvement ideas to other students.

This would be a great opportunity to invite an engineer to visit the class, either physically or using software such as Skype. Tell the engineer ahead of time to make sure to discuss the roles constraint and criteria play in the engineering process.

Online Fossil Models

Topic: Fossils

Grade Levels: 5 and up

Related Standards:

AASL Standards for the 21st-Century Learner

1.1.8 - Demonstrate mastery of technology tools to access information and pursue inquiry.

1.3.5 - Use information technology responsibly.

2.1.4 - Use technology and other information tools to analyze and organize information.

2.1.6 - Use the writing process, media and visual literacy, and technology skills to create products that express new understandings.

3.1.2 - Participate and collaborate as members of a social and intellectual network of learners.

3.1.4 - Use technology and other information tools to organize and display knowledge and understanding in ways that others can view, use, and assess.

3.1.6 - Use information and technology ethically and responsibly.

Common Core Standard

CCSS.ELA.CC.9-10.W.6 - Use technology, including the Internet, to produce, publish, and update individual or shared writing products, taking advantage of technology's capacity to link to other information and to display information flexibly and dynamically.

How to Integrate 3D Modeling and Printing

Resources

http://3d.si.edu/ (Smithsonian X 3D) or

http://africanfossils.org/ (virtual lab with 3D models of fossils that can be 3D printed)

The Smithsonian is one of the first museums to begin to digitize their collection and offer these pieces as part of a 3D model archive that can be 3D printed. After a lesson on characteristics of fossils, introduce students to the Smithsonian Museum. Allow students to research the types of artifacts on display at the Smithsonian museums. Introduce students to the Smithsonian X 3D website and allow them to browse the models and go through the tours available. This research can be integrated into other curriculum areas and research. Although there are not a large number of objects digitized into the Smithsonian X 3D site yet, additional objects are in the process of being added.

The African Fossils site provides 3D models of fossils found around Lake Turkana in East Africa. Hominids, animals, and tool models are available for downloading and 3D printing. Clicking on the photo of each item displays information about when and how that particular fossil was discovered.

Allow students to find a fossil model, research the specimen, and 3D print the fossil from the website.

Molecular Models

Topic: Molecules

Grade Levels: 5 and up

Related Standards:

AASL Standards for the 21st-Century Learner

3.1.4 - Use technology and other information tools to organize and display knowledge and understanding in ways that others can view, use, and assess.

Next Generation Science Standard

5-PS1-1. Develop a model to describe that matter is made of particles too small to be seen.

How to Integrate 3D Modeling and Printing

Molecular models help students to better visualize the shapes of molecules and how those molecules behave. Students will design a model of a molecule, using different colors for different types of atoms. Students can select molecules related to a research topic and design the molecule model using 3D modeling software.

For an extension of this lesson, students can create models of other scientific processes.

Solutions to Local Problems

Topic: Engineering Design

Grade Levels: 6–8

Related Standards:

Next Generation Science Standards

MS-ETS1-1. Define the criteria and constraints of a design problem with sufficient precision to ensure a successful solution, taking into account relevant scientific principles and potential impacts on people and the natural environment that may limit possible solutions.

MS-ETS1-2. Evaluate competing design solutions using a systematic process to determine how well they meet the criteria and constraints of the problem.

MS-ETS1-3. Analyze data from tests to determine similarities and differences among several design solutions to identify the best characteristics of each that can be combined into a new solution to better meet the criteria for success.

MS-ETS1-4. Develop a model to generate data for iterative testing and modification of a proposed object, tool, or process such that an optimal design can be achieved.

How to Integrate 3D Modeling and Printing

Give students a project or scenario with well-defined and detailed criteria and constraints regarding a topic that is a real problem in the local community or that is in current news. If the problem is a local one, have a local engineer that is dealing with the problem visit and share information about the problem with students. Have students research the problem, including what types of solutions have already been tried and how people in the community view the problem or what they would like for a solution. Let students compare possible solutions to determine which of the solutions meet constraint and criteria characteristics, including the values and needs of people in the community.

Students can design objects that solve the problem (scaled if needed) using 3D modeling software. Ensure students go through the engineering process, including modifications, before 3D printing occurs. Students can present models to engineers working on the solution and get feedback on their ideas for solutions.

Cell Models

Topic: Cells. Develop and use a model to describe the function of a cell as a whole and ways parts of cells contribute to the function.

Grade Levels: 6–8

Related Standard:

AASL Standards for the 21st-Century Learner

3.1.4 - Use technology and other information tools to organize and display knowledge and understanding in ways that others can view, use, and assess.

How to Integrate 3D Modeling and Printing

Students will design a model of a cell using 3D modeling software. Cell models can be 3D printed for presentations or displays. An example of an animal cell can be found on Thingiverse.com at thingiverse.com/thing:33360 and an example of a plant cell can be seen at thingiverse.com/thing:439251.

QR Codes

Topic: Connecting with Technology

Grade Levels: 6–12

Related Standards:

AASL **Standards for the 21st-Century Learner**

1.1.7 - Make sense of information gathered from diverse sources by identifying misconceptions, main and supporting ideas, conflicting information, and point of view or bias.

1.2.3 - Demonstrate creativity by using multiple resources and formats.

2.1.6 - Use the writing process, media and visual literacy, and technology skills to create products that express new understandings.

3.1.4 - Use technology and other information tools to organize and display knowledge and understanding in ways that others can view, use, and assess.

How to Integrate 3D Modeling and Printing

QR (Quick Response) Codes have been used for the past few years by teachers and librarians for a variety of purposes. Librarians can combine 3D printing with QR codes to add pizzazz back into QR codes that have become commonplace to many students.

Librarians can provide students with an easy way to connect with the library's website, blog, or webcast by 3D printing a QR code that, when scanned, sends the user automatically to that site.

To 3D print a QR code:

Generate a QR code using a website for generating QR codes. There are many free QR code generating sites that are free, such as:

http://goqr.me

https://www.the-qrcode-generator.com/

http://www.qrstuff.com/

https://qrcode.kaywa.com/

http://www.qr-code-generator.com/

To generate the QR code, type in the website address and then download the QR code that is generated. You can make sure the generated QR code works by scanning it using an app that reads QR codes, such as QR Reader, Scan, QR Code Reader, and i-nigma. There are many QR code reading apps that can be downloaded from the iTunes store or Google Play and they work the same. Try out one or two until you find one you like. To download the generated QR code, click on download and then save the generated code on your computer. This will allow you to retrieve it and work on it using 3D modeling software.

Next, upload the generated QR code into a 3D modeling software program. Because the generated QR code is flat, students will need to outline the shape of the QR code for the program to transform it from a flat shape to a 3D shape. Outlining the shape of the QR code separates the shape from the design grid and allows the program to see the shape as a full 3D shape.

After outlining the QR code to make it 3D, use the scale feature of the 3D modeling software program to enlarge or shrink it to make it the wanted height and width. After scaling the height and width of the QR code, scale the depth of the code.

Next, save the design file in .STL format onto a flash drive or SD card, whichever your printer accepts. Open the 3D printer software and upload the saved design file (saved in .STL format). Use the menu on the 3D printer to find the design file and print.

Use the QR code reader app to ensure the QR code directs users to the desired website.

Ways that 3D printed QR codes can be used include the following:

The librarian can print out QR codes that direct students to the library's website or blog. If possible, attach a key chain and use it as a library marketing tool.

Students can print out QR codes directing others to a student-made website, presentation, or other type of information.

Physics of Flight

Topic: Physics

Grade Levels: 9–12

Related Standards:

AASL Standards for the 21st-Century Learner

1.1.8 - Demonstrate mastery of technology tools to access information and pursue inquiry.

1.3.5 - Use information technology responsibly.

2.1.4 - Use technology and other information tools to analyze and organize information.

2.1.6 - Use the writing process, media and visual literacy, and technology skills to create products that express new understandings.

How to Integrate 3D Modeling and Printing

Students will create physical and conceptual models for planes in flight designed according to the engineering design process.

Resource

http://www.aeronautics.nasa.gov/mib.htm

NASA's Museum in a Box provides students and educators with "hands-on, minds-on" lessons about the physical science of flight. Educators can utilize these lessons to have students create objects, which can then be 3D printed. This would be a great way to end a unit on the physics of flight or as an extra project for students.

Global Challenges

Topic: Engineering Design

Grade Levels: 9–12

Related Standards:

Next Generation Science Standards

HS-ETS1-1. Analyze a major global challenge to specify qualitative and quantitative criteria and constraints for solutions that account for societal needs and wants.

HS- HS-ETS1-1. Analyze a major global challenge to specify qualitative and quantitative criteria and constraints for solutions that account for societal needs and wants.

HS-ETS1-2. Design a solution to a complex real-world problem by breaking it down into smaller, more manageable problems that can be solved through engineering.

HS-ETS1-3. Evaluate a solution to a complex real-world problem based on prioritized criteria and trade-offs that account for a range of constraints, including cost, safety, reliability, and aesthetics, as well as possible social, cultural, and environmental impacts.

HS-ETS1-4. Use a computer simulation to model the impact of proposed solutions to a complex real-world problem with numerous criteria and constraints on interactions within and between systems relevant to the problem.

How to Integrate 3D Modeling and Printing

Have students research and consider some of the major global challenges in the world today. Allow students to research and choose a problem that is meaningful to them. Possible global challenges students could choose to research include disease, pollution, lack of clean water, lack of food, and war. Allow students time to discuss which of these global challenges could be addressed through engineering. This may take time as students consider these global problems and what can be done to fix them. Have students write down possible engineering solutions to these problems and choose one to address with their own engineering. Students should consider constraints and criteria in the real world, such as safety, cost, and cultural impacts of the possible solutions. Have students consider if 3D printing might be part of the solution for solving some of these global problems.

Students can design solutions in 3D modeling software and then print scaled models or can create their own solutions.

11

3D Modeling and Printing in the Language Arts Curriculum

It may seem more natural for 3D modeling and printing to be integrated into the Math and Science curriculum than into the Language Arts curriculum. The reality is that 3D modeling and printing can easily be a part of the Language Arts curriculum, including reading and writing. These processes provide opportunities for engaging those students who have checked out of Language Arts because they struggle with reading or writing. Engaging them through a different medium may be just what they need to connect with the curriculum.

Increasing Vocabulary

Topic: College and Career Readiness—Reading

Grade Levels: K–12

Related Standards:

AASL Standards for the 21st-Century Learner

1.1.4 - Find, evaluate, and select appropriate sources to answer questions.

1.1.6 - Read, view, and listen for information presented in any format (e.g., textual, visual, media, digital) in order to make inferences and gather meaning.

2.2.1 - Demonstrate flexibility in use of resources by adapting information strategies to each specific resource and by seeking additional resources when clear conclusions cannot be drawn.

Common Core Standard

CCSS.ELA-LITERACY.CCRA.R.4 - Interpret words and phrases as they are used in a text, including determining technical, connotative, and figurative meanings, and analyze how specific word choices shape meaning or tone. Students apply a wide range of strategies to comprehend, interpret, evaluate, and appreciate texts.

How to Integrate 3D Modeling and Printing

The ELA Common Core Standards focus on reading complex text closely and intentionally. Reading closely and studying word use help students understand deeper meanings and evaluate text. Students often have difficulty learning new vocabulary words and 3D printing is a way to make a connection between words and tangible objects.

Use 3D modeling and printing to teach denotation: Give students a list of vocabulary words. Students can look up meanings and write meanings in their own words. Students use 3D modeling software to design an object that represents the new word or in some way defines that word. The librarian or students can 3D print a representation for each new word or can choose objects to be printed.

Use 3D modeling and printing to teach connotation: Give students a list of vocabulary words. Discuss the meanings of each word and then discuss the feelings and ideas invoked by each word. Students then design an object using 3D modeling software that represents each word's connotation. The librarian or students can 3D print each representative model or can choose which objects will be printed. Librarians could use the two lesson ideas above with the same vocabulary words, teaching first denotation and then connotation, giving students a full understanding of the new words.

Analyzing Poetry

Topic: College and Career Readiness—Reading

Grade Levels: K–12

Related Standards:

AASL **Standards for the 21st-Century Learner**

1.1.4 - Find, evaluate, and select appropriate sources to answer questions.

1.1.6 - Read, view, and listen for information presented in any format (e.g., textual, visual, media, digital) in order to make inferences and gather meaning.

2.2.1 - Demonstrate flexibility in use of resources by adapting information strategies to each specific resource and by seeking additional resources when clear conclusions cannot be drawn.

Common Core Standard

CCSS.ELA-LITERACY.CCRA.R.4 - Interpret words and phrases as they are used in a text, including determining technical, connotative, and figurative meanings, and analyze how specific word choices shape meaning or tone. Students apply a wide range of strategies to comprehend, interpret, evaluate, and appreciate texts.

How to Integrate 3D Modeling and Printing

Analyzing poetry is a great way for students to interpret words and meanings. Read poetry with students or allow students to write their own poetry. Students can design objects in 3D modeling software that represents the theme or feeling of the poem.

Evaluating Sources

Topic: College and Career Readiness—Reading

Grade Levels: K–12

Related Standards:

AASL Standards for the 21st-Century Learner

1.1.1 - Follow an inquiry-based process in seeking knowledge in curricular subjects and make the real world connection for using this process in own life.

1.1.2 - Use prior and background knowledge as context for new learning.

1.1.4 - Find, evaluate, and select appropriate sources to answer questions.

1.1.6 - Read, view, and listen for information presented in any format (e.g., textual, visual, media, digital) in order to make inferences and gather meaning.

1.3.1 - Respect copyright/intellectual property rights of creators and producers.

1.3.3 - Follow ethical and legal guidelines in gathering and using information.

2.1.2 - Organize knowledge so that it is useful.

Common Core Standard

CCSS.ELA-LITERACY.CCRA.R.7 - Integrate and evaluate content presented in diverse media and formats, including visually and quantitatively, as well as in words.

How to Integrate 3D Modeling and Printing

One of the most important skills students can learn is to evaluate the information they find during research. Every school librarian knows that many students lean toward the "if it's on the Internet, it has to be real" way of thinking. Teaching students to evaluate the source of the information and to determine any bias that may be present can be a challenging task.

Librarians can use technology topics, and 3D modeling and printing specifically, to engage students in research and teach them how to become wise consumers of information. Have students research 3D modeling and printing on sites that are from expert sources on the topic. Next, have students review websites that are more informal sources, such as blogs, tweets, and YouTube videos about 3D modeling and printing. Have students compare the types of information provided from each type of source to determine which is more reliable for gathering information.

Visual Models

Topic: Presentation of Knowledge and Ideas

Grade Levels: K–12

Related Standards:

AASL Standards for the 21st-Century Learner

1.2.3 - Demonstrate creativity by using multiple resources and formats.

2.1.6 - Use the writing process, media and visual literacy, and technology skills to create products that express new understandings.

3.1.3 - Use writing and speaking skills to communicate new understandings effectively.

3.1.4 - Use technology and other information tools to organize and display knowledge and understanding in ways that others can view, use, and assess.

Common Core Standard

CCSS.ELA-LITERACY.CCRA.SL.5 - Make strategic use of digital media and visual displays of data to express information and enhance understanding of presentations.

How to Integrate 3D Modeling and Printing

3D modeling and printing is an effective way for students to present information in a visual manner that is much different than drawings or posters. Anytime students must present information visually, allow students to create using 3D modeling software and then 3D print the object(s). Students will start recognizing 3D modeling and printing as a viable option for presenting information and utilize it along with all of the other methods of presenting information they have at their disposal.

Prompts You Can Hold in Your Hand

Topic: Writing

Grade Levels: K–12

Related Standards:

AASL Standards for the 21st-Century Learner

2.1.1 - Continue an inquiry-based research process by applying critical thinking skills (analysis, synthesis, evaluation, organization) to information and knowledge in order to construct new understandings, draw conclusions, and create new knowledge.

2.1.4 - Use technology and other information tools to analyze and organize information.

3.1.4 - Use technology and other information tools to organize and display knowledge and understanding in ways that others can view, use, and assess.

4.3.2 - Recognize that resources are created for a variety of purposes.

Common Core Standards

Production and Distribution of Writing:

CCSS.ELA-LITERACY.W.4 - Produce clear and coherent writing in which the development and organization are appropriate to task, purpose, and audience.

CCSS.ELA-LITERACY.W.5 - With guidance and support from peers and adults, develop and strengthen writing as needed by planning, revising, and editing.

CCSS.ELA-LITERACY.W.6 - With some guidance and support from adults, use technology, including the Internet, to produce and publish writing as well as to interact and collaborate with others; demonstrate sufficient command of keyboarding skills to type a minimum of one page in a single sitting.

Range of Writing:

CCSS.ELA-LITERACY.W.10 - Write routinely over extended time frames (time for research, reflection, and revision) and shorter time frames (a single sitting or a day or two) for a range of discipline-specific tasks, purposes, and audiences.

How to Integrate 3D Modeling and Printing

The librarian will use 3D modeling software to design an object or can find a model that has already been created on a site such as Thingiverse.com. Print the object on a 3D printer and use the object for a writing prompt. Display the object in a place that all students are able to see it or pass the object around so students can see details. Students will create a story involving that object and submit it to the library website or blog. The librarian will upload a picture of the object to the site so readers can see what the object was that was being written about.

Book Covers

Topic: Reading

Grade Levels: K–12

Related Standards:

AASL **Standards for the 21st-Century Learner**

3.1.4 - Use technology and other information tools to organize and display knowledge and understanding in ways that others can view, use, and assess.

Common Core Standards

CCSS.ELA-LITERACY.SL.9-10.5 - Make strategic use of digital media (e.g., textual, graphical, audio, visual, and interactive elements) in presentations to enhance understanding of findings, reasoning, and evidence and to add interest.

CCSS.ELA-LITERACY.SL.11-12.5 - Make strategic use of digital media (e.g., textual, graphical, audio, visual, and interactive elements) in presentations to enhance understanding of findings, reasoning, and evidence and to add interest.

How to Integrate 3D Modeling and Printing

As part of a novel study, students will design a front cover for the book using 3D modeling software. Covers will be 3D printed and displayed or presented by students.

High-Frequency Words

Topic: Phonics and Word Recognition

Grade Level: K

Related Standards:

AASL Standards for the 21st-Century Learner

1.1.2 - Use prior and background knowledge as context for new learning.

Common Core Standards

CCSS.ELA-LITERACY.RF.K.3 - Know and apply grade-level phonics and word analysis skills in decoding words.

CCSS.ELA-LITERACY.RF.K.3.C - Read common high-frequency words by sight (e.g., *the*, *of*, *to*, *you*, *she*, *my*, *is*, *are*, *do*, *does*).

How to Integrate 3D Modeling and Printing

Students will use 3D modeling software to create words that will be 3D printed. Most software programs allow users to easily design words. If help is needed, search sites, such as Instructables.com, for step-by-step directions on 3D printing text using the chosen 3D modeling software. After 3D printing words created by each student, words can be used during group word work or by individual students. Students love to use these printed words to quiz each other. The librarian can share these with the classroom teacher, who can carry these in a bucket and quiz students as they are waiting in line for the restroom or for lunch.

Analyzing Characters, Settings, and Events

Topic: Key Ideas and Details

Grade Level: 1

Related Standards:

AASL Standards for the 21st-Century Learner

4.1.3 - Respond to literature and creative expressions of ideas in various formats and genres.

Common Core Standard

CCSS.ELA-LITERACY.RL.1.3 - Describe characters, settings, and major events in a story, using key details.

How to Integrate 3D Modeling and Printing

The librarian should provide students a lesson on analyzing characters, settings, and events in a story. Allow students to select a character, setting, or event to describe in detail. It would be helpful to allow students to write down the details and use their written notes as they design an object that represents their chosen character, setting, or event. Students will then present their objects to other students, explaining key details as part of their presentations.

An alternative is to have students or groups of students design and print different components of a story. After all objects are printed, students (individually or in groups) can present the story using their 3D printed components.

Prefixes and Suffixes

Topic: Phonics and Word Recognition

Grade Level: 3

Related Standards:

AASL Standards for the 21st-Century Learner

4.1.3 - Respond to literature and creative expressions of ideas in various formats and genres.

Common Core Standards

CCSS.ELA-LITERACY.RF.3.3 - Know and apply grade-level phonics and word analysis skills in decoding words.

CCSS.ELA-LITERACY.RF.3.3.A - Identify and know the meaning of the most common prefixes and derivational suffixes.

How to Integrate 3D Modeling and Printing

Students will design prefixes and suffixes in 3D design software. These will be 3D printed and used in the classroom to teach the meanings of each.

Themes

Topic: Key Ideas and Details—State Award Book Lists

Grade Level: 4

Related Standards:

AASL Standards for the 21st-Century Learner

1.1.6 - Read, view, and listen for information presented in any format (e.g., textual, visual, media, digital) in order to make inferences and gather meaning.

Common Core Standard

CCSS.ELA-LITERACY.RL.4.2 - Determine a theme of a story, drama, or poem from details in the text; summarize the text.

How to Integrate 3D Modeling and Printing

3D modeling and printing allows students to move from an abstract concept or theme of a story to a concrete object that represents that concept or theme. After reading a story, drama, or poem, have students analyze the characters, action, and setting to determine the theme of the text. I tell students they have to be text detectives when determining the theme. Students should be comfortable with evaluating the title, repeating patterns or symbols, and allusions among other factors to determine the theme. After students work individually or in groups to decide the theme of the text, they will use 3D modeling software to design an object that represents the themes. For a challenge, have students work together to make one object that represents all of the themes of the story, drama, or poem.

Extension: As the librarian promotes or reads their state book award list, Caldecott Medal, Newbery Medal, Pura Belpré, or Coretta Scott King (or any other book list) winners with students, discuss the theme of each book. The library could sponsor a contest in which students choose a book from one of the above lists and design an object that represents that book using 3D modeling software. Objects could be 3D printed and displayed in the library. If part of a contest, students could promote their chosen book with the 3D printed object during morning announcements or at an event. The librarian could have school administrators, teachers, or parents judge the entries and select winners. Even without the contest, this would be a nice activity to promote 3D modeling and printing within the student body and to promote the library in the process.

Mythological Allusions

Topic: Craft and Structure

Grade Level: 4

Related Standards:

AASL Standards for the 21st-Century Learner

1.1.6 - Read, view, and listen for information presented in any format (e.g., textual, visual, media, digital) in order to make inferences and gather meaning.

Common Core Standard

CCSS.ELA-LITERACY.RL.4.4 - Determine the meaning of words and phrases as they are used in a text, including those that allude to significant characters found in mythology (e.g., Herculean).

How to Integrate 3D Modeling and Printing

Allusions to mythological characters occur often in literature. Some of the most common allusions include "a Herculean effort," "opening Pandora's Box," "having a Midas touch," "Achilles' heel," "mercurial," and "chronology." Students can research the meaning of the allusions and work in small groups to design objects in 3D modeling software that represent the allusions to mythological characters and then 3D print the objects.

Compare and Contrast Characters, Settings, and Events

Topic: Key Ideas and Details

Grade Level: 5

Related Standards:

AASL Standards for the 21st-Century Learner

1.1.6 - Read, view, and listen for information presented in any format (e.g., textual, visual, media, digital) in order to make inferences and gather meaning.

Common Core Standard

CCSS.ELA-LITERACY.RL.5.3 - Compare and contrast two or more characters, settings, or events in a story or drama, drawing on specific details in the text (e.g., how characters interact).

How to Integrate 3D Modeling and Printing

Students will use 3D modeling software to design an object that represents the characters, settings, or events in a story. The librarian can assign characters, settings, or events to small groups of students to ensure a variety. Students will use the printed objects to compare and contrast the printed objects that represent characters, settings, or events.

Storyboarding

Topic: Key Ideas and Details

Grade Level: 6

Related Standards:

AASL Standards for the 21st-Century Learner

3.1.3 - Use writing and speaking skills to communicate new understandings effectively.

Common Core Standard

CCSS.ELA-LITERACY.RL.6.3 - Describe how a particular story's or drama's plot unfolds in a series of episodes as well as how the characters respond or change as the plot moves toward a resolution.

How to Integrate 3D Modeling and Printing

After reading a story, have students storyboard the entire story and include information about how characters responded or changed in the series of events. Students can use a simple storyboard template, such as the one in the figure to the right.

Students will lead discussions on their storyboards and how characters changed throughout the story. Students then will use 3D modeling software to design the character or they can design an object that represents the change the character went through.

Writing a 3D Printing Guide for Students

Topic: Key Ideas and Details

Grade Levels: 6–8

Related Standards:

AASL Standards for the 21st-Century Learner

1.1.1 - Follow an inquiry-based process in seeking knowledge in curricular subjects and make the real world connection for using this process in own life.

1.2.1 - Display initiative and engagement by posing questions and investigating the answers beyond the collection of superficial facts.

1.2.5 - Demonstrate adaptability by changing the inquiry focus, questions, resources, or strategies when necessary to achieve success.

1.4.1 - Monitor own information seeking processes for effectiveness and progress, and adapt as necessary.

1.4.3 - Monitor gathered information and assess for gaps or weaknesses.

Common Core Standard

CCSS.ELA-LITERACY.RST.6-8.3 - Follow precisely a multistep procedure when carrying out experiments, taking measurements, or performing technical tasks.

How to Integrate 3D Modeling and Printing

This lesson can be applied to a variety of 3D modeling and printing projects. Students must follow multistep procedures every time they utilize 3D modeling software and the 3D printer. To integrate this with writing, have students write a guide for other students who are new to 3D modeling and printing. This guide can be shared with new students as they come in during the year to help them learn the 3D modeling process.

Presentation Props

Topic: Presentation of Knowledge and Ideas

Grade Levels: 9–12

Related Standards:

AASL Standards for the 21st-Century Learner

2.1.6 - Use the writing process, media and visual literacy, and technology skills to create products that express new understandings.

3.1.3 - Use writing and speaking skills to communicate new understandings effectively.

3.3.4 - Create products that apply to authentic, real-world contexts.

Common Core Standards

CCSS.ELA-LITERACY.SL.9-10.5 - Make strategic use of digital media (e.g., textual, graphical, audio, visual, and interactive elements) in presentations to enhance understanding of findings, reasoning, and evidence and to add interest.

CCSS.ELA-LITERACY.SL.11-12.5 - Make strategic use of digital media (e.g., textual, graphical, audio, visual, and interactive elements) in presentations to enhance understanding of findings, reasoning, and evidence and to add interest.

How to Integrate 3D Modeling and Printing

Students will utilize 3D printed objects during presentations to not only add interest, but to truly personalize the presentation experience. Because students design their own models to utilize as part of their presentation, this promotes creativity and individuality.

12

3D Modeling and Printing in the Social Studies Curriculum

In recent years, the social studies curriculum has taken a backseat to STEM curriculum areas. 3D modeling and printing is a great way to integrate STEM into the Social Studies curriculum. Lessons presented in this chapter cover a range of grade levels and topics. Fun ways to introduce or reinforce vocabulary, ideas for connecting students with historical characters and events, and mixing writing with 3D modeling and printing are just a few of the lesson ideas presented. Utilizing 3D modeling and printing in the library will allow librarians to revitalize the Social Studies curriculum and create excitement.

Primary Sources

Topic: Key Ideas and Details

Grade Levels: K–12

Related Standards:

AASL Standards for the 21st-Century Learner

2.1.1 - Continue an inquiry-based research process by applying critical thinking skills (analysis, synthesis, evaluation, organization) to information and knowledge in order to construct new understandings, draw conclusions, and create new knowledge.

2.1.3 - Use strategies to draw conclusions from information and apply knowledge to curricular areas, real world situations, and further investigations.

Common Core Standard

CCSS.ELA-LITERACY.RH.6-8.2 - Determine the central ideas or information of a primary or secondary source; provide an accurate summary of the source distinct from prior knowledge or opinions.

How to Integrate 3D Modeling and Printing

Primary sources can be shared with students, regardless of grade level. For younger students, photographs and audio recordings are both simple but effective ways to share primary sources. Photographs and audio recordings will also engage older students.

One of the best resources available for 3D printing sources is the Smithsonian Institute. The Smithsonian Institute is providing educators and students with a way to access the Smithsonian's collection that the public doesn't see. The Smithsonian Digitization Project Office is working on the vast undertaking of digitizing objects from all 19 Smithsonian museums, 9 research centers, as well as the National Zoo. The Smithsonian's total collection is over 137 million objects and although digitizing them all would take well over 200 years, they are attempting to digitize at least 10 percent of their collection for the public to access through their digitization website. Objects from the Smithsonian's collection can be accessed through the website, http://3d.si.edu/. This site provides detailed tours of each object, such as the 1903 Wright Flyer, and includes photographs and text explaining what is being shown in the 3D tour. Objects that can be downloaded and 3D printed include the Wright Flyer, Amelia Earhart's flight suit, and life masks of Abraham Lincoln. To view models that are available for 3D printing, go to http://3d.si.edu/ and click on Browse Models. To see available tours, click on Browse Tours.

Vocabulary

Topic: Craft and Structure

Grade Levels: 6–8

Related Standards:

AASL Standards for the 21st-Century Learner

4.1.3 - Respond to literature and creative expressions of ideas in various formats and genres.

Common Core Standard

CCSS.ELA-LITERACY.RH.6-8.4 - Determine the meaning of words and phrases as they are used in a text, including vocabulary specific to domains related to history/social studies.

How to Integrate 3D Modeling and Printing

As part of teaching students words or phrases that are new to them, have students design objects in a modeling software program that represents those words or phrases. Students will 3D print tangible objects that represent new vocabulary words to help increase understanding of the new vocabulary words.

Models

Topic: Integration of Knowledge and Ideas

Grade Levels: 6–8

Related Standards:

AASL Standards for the 21st-Century Learner

1.1.7 - Make sense of information gathered from diverse sources by identifying misconceptions, main and supporting ideas, conflicting information, and point of view or bias.

1.2.3 - Demonstrate creativity by using multiple resources and formats.

2.1.6 - Use the writing process, media and visual literacy, and technology skills to create products that express new understandings.

3.1.4 - Use technology and other information tools to organize and display knowledge and understanding in ways that others can view, use, and assess.

Common Core Standard

CCSS.ELA-LITERACY.RH.6-8.7 - Integrate visual information (e.g., in charts, graphs, photographs, videos, or maps) with other information in print and digital texts.

How to Integrate 3D Modeling and Printing

Librarians will make learning more visual by either designing an object in 3D modeling software or by finding a model already created and just printing it. Using these visuals allows students to see objects in history as they were meant to be seen, and not just on a computer screen. As students become more adept at 3D modeling and printing, the librarian can turn the actual modeling part over to students. Students can design and print an object to share with their class, other classes in the district, or with a partner class anywhere around the world using Skype or FaceTime.

Events and Artifacts

Topic: Historical Artifacts

Grade Levels: 6–12

Related Standards:

AASL **Standards for the 21st-Century Learner**

1.1.7 - Make sense of information gathered from diverse sources by identifying misconceptions, main and supporting ideas, conflicting information, and point of view or bias.

1.2.3 - Demonstrate creativity by using multiple resources and formats.

2.1.6 - Use the writing process, media and visual literacy, and technology skills to create products that express new understandings.

3.1.4 - Use technology and other information tools to organize and display knowledge and understanding in ways that others can view, use, and assess.

How to Integrate 3D Modeling and Printing

Students research a historical event with a focus on artifacts that could be associated with that event. Students then use 3D design software to design artifacts that represent or are associated with the event. After 3D printing the artifacts, students will create videos presenting information on the historical event and artifact.

Historical Characters

Topic: Important People in History

Grade Levels: 6–12

Related Standards:

AASL Standards for the 21st-Century Learner

1.1.7 - Make sense of information gathered from diverse sources by identifying misconceptions, main and supporting ideas, conflicting information, and point of view or bias.

1.2.3 - Demonstrate creativity by using multiple resources and formats.

2.1.6 - Use the writing process, media and visual literacy, and technology skills to create products that express new understandings.

3.1.4 - Use technology and other information tools to organize and display knowledge and understanding in ways that others can view, use, and assess.

How to Integrate 3D Modeling and Printing

Students use 3D design software to design historical characters using 3D modeling software. Students then print the historical characters on a 3D printer. The librarian can create a blog if they don't already have one and have students write a blog entry from that person's viewpoint regarding an important event that occurred during that person's lifetime.

Three Things

Topic: Representations of History

Grade Levels: 6–12

Related Standards:

AASL **Standards for the 21st-Century Learner**

1.1.7 - Make sense of information gathered from diverse sources by identifying misconceptions, main and supporting ideas, conflicting information, and point of view or bias.

1.2.3 - Demonstrate creativity by using multiple resources and formats.

2.1.6 - Use the writing process, media and visual literacy, and technology skills to create products that express new understandings.

3.1.4 - Use technology and other information tools to organize and display knowledge and understanding in ways that others can view, use, and assess.

How to Integrate 3D Modeling and Printing

Divide students into small groups and challenge each group to use 3D design software to create three objects that represent a historical event. Each group then presents to the rest of the class about the significance of the three items they chose to design and print.

Geocaching

Topic: Geography

Grade Levels: 6–12

Related Standards:

AASL Standards for the 21st-Century Learner

1.2.3 - Demonstrate creativity by using multiple resources and formats.

3.1.4 - Use technology and other information tools to organize and display knowledge and understanding in ways that others can view, use, and assess.

How to Integrate 3D Modeling and Printing

Geocaching involves hiding and finding hidden objects using GPS coordinates. Librarians can combine 3D modeling and printing with geocaching for a fun way to engage students that just may get students hooked on geocaching. Librarians can start a geocaching club for students to continue the fun.

After introducing the idea of geocaching to students, have them visit www.geocaching.com to learn more about geocaching. The site has much information for those new to geocaching under the Learn tab, called Geocaching 101. The site also contains videos on how to find and hide a geocache, geocaching etiquette, and even includes a video explaining geocaching in Spanish.

Allow students to design an object in 3D modeling software they would like to hide for others to find. After printing, students can hide the objects around the campus and can post coordinates for others to find the objects.

Historical Markers

Topic: Commemorating Historical Events

Grade Levels: 9–12

Related Standards:

AASL Standards for the 21st-Century Learner

1.1.7 - Make sense of information gathered from diverse sources by identifying misconceptions, main and supporting ideas, conflicting information, and point of view or bias.

1.2.3 - Demonstrate creativity by using multiple resources and formats.

2.1.6 - Use the writing process, media and visual literacy, and technology skills to create products that express new understandings.

3.1.4 - Use technology and other information tools to organize and display knowledge and understanding in ways that others can view, use, and assess.

How to Integrate 3D Modeling and Printing

Open this activity with a discussion of ways historical events and places are commemorated in the United States. Show examples of historical markers around your city or state or throughout the United States. Make sure to show examples of different types of historical markers, such as plaques and roadside historical markers. Have students select an event or location for which they would like to design a historical marker. Students can work individually or in groups to research their chosen location or event. Discuss with students how the information needs to be evaluated to determine what information should be included on the historical marker. Have students design the historical marker on paper. After designing on paper, students will utilize 3D modeling software to design the historical marker and then 3D print the design.

After printing each historical marker, students can research the process for getting a historical marker placed at a location in your state. Most of the forms for requesting a historical marker can be found online and require requestors to explain the significance of the location and the rationale for why the particular location deserves to have a historical marker. The librarian can make up a mock form or use the real form for students to complete and submit to the librarian. As an extension, students can present their rationales for the markers to other students, teachers, or parents.

13

Using 3D Modeling and Printing in the Fine Arts Curriculum

Often, fine arts teachers are the first on their campus to become interested in 3D modeling and printing. A librarian looking for a teacher to partner with to begin learning 3D modeling and printing should consider collaborating with a fine arts teacher to write a grant or to purchase the printer. The arts and 3D modeling and printing create a natural partnership that interests students of all ages.

3D modeling and printing can be used in a variety of ways in an art, music, or theater arts classroom or program. Fine arts programs encourage creativity and self-expression, making these programs a perfect avenue for using 3D modeling and printing.

Theater Arts

In a theater arts or drama program, students can design and print props for theater performances. Imagine students holding props that started as ideas in their own heads. Students could design the prop as they envision it should be, strategically choosing the color of filament to use to print it. 3D printing allows the creation of unique or unusual costume or set pieces that can't be purchased or are too expensive to purchase.

Owen Collins, a professor of Theater at Washington and Lee University, utilizes 3D printing as a method for creating scaled models of stage sets. This allows sets to be designed inexpensively in a scaled model and changed as needed before creating the final design. In his article "Affordable 3D Printing," Collins (2012) notes that creating these scaled models allows him to easily see the spatial relationships between pieces during the set design process. Students can use this same idea to build scaled pieces for stage sets.

Art

3D art is becoming popular as artists discover 3D printed objects as a new medium for expression. Artists are using 3D printing in interior design, sculpture, and to replicate classical art in a new way. 3D printing is also being used by artists to help restore pieces that are too fragile to manipulate. The original piece is scanned, scaled, and 3D printed to allow artists the ability to determine the best way to restore the piece without damaging it. School librarians can promote 3D artistry by providing information or a display on artists who use 3D printing as a tool. Students will be inspired to design their own pieces of art. The librarian can provide student artists with space for an art show where they can display their pieces of art.

3D modeling and printing is a great way to teach students about architectural design. After learning about characteristics of different styles of architecture, students can use 3D modeling software to design their own structure with the same characteristics. Students can also design their version of architecture from around the world, or design structures that would be suitable for humans if they were able to live on other planets.

Often, art in many forms is found at archaeological dig sites. Archeologists are using photomodeling, a process in which individual fragments of an ancient sculpture are photographed from many different angles to create 3D models of each piece. The individual fragments are then pieced together to form a 3D model of the original artifact. This creates a model that is a good representation of the original that can be evaluated and manipulated. A team at Cornell has successfully 3D printed cuneiform tablets that allow experts to interact with the tablets in a way never thought possible. Links to the cuneiform models can be found on the website and printed so anyone can have access to these ancient tablets from ancient Mesopotamia.

http://creativemachines.cornell.edu/cuneiform

Along these same lines, students can design their own versions of archaeological artifacts using 3D modeling software and then print them. In the future, more museums and universities with artifact holdings will hopefully make 3D models of their holdings available to the public so students can access and 3D print them. In the future, the majority of museum holdings will be available to the general public anywhere in the world. Models will be accessed through the museum website and will be able to be downloaded and 3D printed.

Music

A fun way to incorporate 3D printing into a music program is to have students use 3D modeling software to design musical instruments they can 3D print and play. Models that are already created can be accessed for free on websites such as Thingiverse.com by searching for "musical instrument." Models for just about any type of musical instrument can be found, including guitars, flutes, saxophones, violins, drum kits, and keyboards have all been successfully 3D printed and used. Lund University's Malmö Academy of Music in Sweden recently held a concert where all instruments were 3D printed. A popular model on Thingiverse.com is for the Makerlele, a 3D printed ukulele (http://www.thingiverse.com/thing:34363). After printing, a student would only need to add bolts and strings to have a fully functional ukulele that can be played. Individual pieces of instruments can also be printed, such as a mouthpiece that can be custom made for the musician.

Resources for 3D Modeling and Printing in the Arts

3D Print Show

3Dprintshow.com

3D Print Show is a comprehensive website that covers a variety of 3D printing topics, but has a strong focus on using 3D printing in design and art. 3D Print Show hosts several 3D shows around the world and features photos and information on objects from the shows. This site gives students concrete examples of how 3D printing is being used in the arts. Concerts using 3D printed instruments and a fashion show where models wear 3D printed fashions are intriguing features on the site.

Josh Harker, Artist

Joshharker.com

Josh Harker is one of the premier artists using 3D printing as a medium for art and sculpture. Students will be amazed and inspired by Harker's intricate 3D printed designs.

Gilles Azzaro, Artist

gillesazzaro.com

Gilles Azzaro is an innovative artist who is combining soundwaves and 3D printing to create sculptures. Azzaro takes sound recordings, analyzes the wavelengths, and preserves them using 3D printing.

14

Tips for Successfully Integrating 3D Modeling and Printing into the Curriculum

Incorporate 3D Modeling and Printing into Authentic Learning Experiences

Authentic learning is a style of learning that encourages students to connect with real-world issues and problems. These types of experiences provide students with meaningful learning of the processes and skills they can take outside the school setting. Authentic learning consists of students working on a genuine problem and taking on the role of a professional. Authentic learning can occur at all levels and with all ages of students but it will look different depending on the ages of the students.

3D modeling and printing can easily be incorporated into problem- or inquiry-based learning. Providing students with the opportunity to research and determine a real-world problem or challenge they can solve makes this technology a powerful tool for any librarian. Students enjoy inquiry- and problem-based learning because connecting what they are doing with real-world problems makes them feel like they are part of something larger. Students around the world are using 3D modeling and printing to learn production development. Students in California were challenged to design a lid for a stainless steel water bottle. The students not only designed a lid for the bottle, but their design became the prototype for the actual lid production. These same students have also been involved in the design and production of several other products that are sold worldwide.

Integrate 3D Modeling and Printing as Part of a STEM or STEAM Initiative

Integrating STEM or STEAM into school libraries has been getting more attention in the past few years. School libraries are providing students with more than just books. Librarians have integrated language arts into the library curriculum but are now seeing the benefits of also integrating science, technology, engineering, arts, and math into the library. Not only does expanding our reach benefit students and support what is being taught in the classroom, but doing so also provides librarians with another opportunity to prove the value of school libraries.

Many school libraries are integrating STEM by adding a makerspace. A makerspace can be any space in which students can create, tinker, and collaborate. It doesn't have to be a dedicated space in the library if there isn't room for that; it can simply be a cart or table with supplies. A 3D printer can easily be incorporated into a makerspace and be utilized to teach science, technology, engineering, and math. Having a 3D printer and computers available for students to access 3D design software will enable students to engage in the engineering process from beginning to end.

Use 3D Modeling and Printing to Teach the Engineering Design Process

The Engineering Design Process is a series of steps engineers use to guide them in problem solving. Engineers must ask a question, imagine a solution, plan a design, create the model, experiment and test that model, then take time to improve the original.

The Engineering Design Process

Step 1: Ask a Question: What exactly is the problem? What are some ways that others have tried to solve the problem?

Step 2: Imagine a Solution: What are some possible solutions? Brainstorm possible solutions with others and choose the best solution.

Step 3: Plan a Design: What does the solution look like? Draw a diagram showing dimensions of the object.

Step 4: Create the Model: Follow the plan created and actually make the object.

Step 5: Test the Model and Make Improvements: Does the model work? What would make it work better?

Students should get the idea that the Engineering Design Process is not only completed one time. The process should be repeated as many times as needed. This may be difficult for some students who are ready to give up after the first time through the process. It may be helpful to connect with professional engineers so students can hear directly from them the importance of redesign in the process. The benefit of teaching the Engineering Design Process to students is that this process can be used for all types of problem solving, not only for engineering.

Include Teachers, Staff, Parents, and the Community

Many school librarians are interested in adding 3D modeling and printing as a curriculum tool but aren't sure they have the knowledge to do so. One of the ideas behind the maker mentality is that of the librarian as a learner themselves. The librarian doesn't have to be the one who knows it all and then teaches it to students.

One of the most important things a school librarian can do after getting a 3D printer is to present a brief training session on the 3D printer to teachers and staff. An easy way to do this is to give a short presentation to staff as part of a staff meeting. Many educators have read articles about 3D printing but have never seen one in action. Have an object printing throughout the day in the library will guarantee staff will stop and ask about it. Providing teachers and staff with information about 3D printing and encouraging them to spend some time designing and printing will give them an opportunity to learn about the 3D printing process and they in turn will be more likely to think of how they can utilize 3D design and printing in the curriculum they teach. Often, there will be at least one teacher in the building who has an interest in integrating 3D printing into the curriculum they are teaching. Finding this person and capitalizing on their interest in 3D printing will build interest from other teachers when they begin to see the excitement from students.

Send a note home to parents and ask for help. Many parents would be interested and willing to help, even if it means they would be learning the 3D modeling and printing process themselves. There may be parents who are familiar with or who use 3D printing in their careers and would be willing to share their knowledge with students. Imagine parents' excitement when they are able to witness first hand their child's excitement for learning and learning.

Search out community members with experience in 3D printing who are willing to help out. Many companies provide employees with time off to volunteer, and these individuals would love to spend time with students, whether it is for a one-time visit or on a regular basis. Research industries using 3D printing, such as the automotive industry, and ask someone to make a visit or to Skype in to show students some examples of how they use 3D printing.

Promoting 3D Printing through Clubs, Open Houses, Maker Fairs, and EdCamps

A simple and easy way to integrate 3D printing into the school library is through a club or group that meets on a regular basis. This not only helps simplify matters for the school librarian's schedule, but also allows students to learn 3D design and printing in depth. Students who use the design software and printer regularly will truly learn how it works, will move beyond creating simple designs, and will be able to challenge themselves with more complicated designs.

Promoting the 3D printer as part of what the school library offers to students and families can be a great way to show the value and relevance of the library while providing students and their family greater access to new and exciting technology. Open the library during open house or other schoolwide events and allow students and their parents to design an object that can be printed. Students can design an object and then save it using their name and/or student identification number. This will allow the librarian to print the object and create a display with student names to showcase student projects. Remember to include teachers and staff-created objects in the display as well.

A mini maker faire sponsored by the school library is a great way for a school or district to promote the maker movement and build excitement for 3D printing within the community. Student groups, parents and families, staff, community members, and local maker groups can bring supplies and materials and assist students in tinkering, creating, and building. The school librarian could include 3D design and printing as part of the mini maker faire and have an object printing during the faire. Nothing draws a crowd like a 3D printer in action. Adults and children alike become mesmerized by the pied piper that is 3D printing.

EdCamps are an effective way to have librarians and teachers come together and teach each other about technology. EdCamps are great ways to provide and receive professional learning on a variety of technology topics. There is no schedule at an EdCamp, and instead, participants sign up for what they would like to teach when they arrive to the EdCamp. Other participants then sign up for the sessions they would like to attend. Sessions on 3D modeling and printing are popular, especially if training on the topic is not widely available in the area.

Leave Quiet Behind

Gone are the days of silence and shushing in the school library. To generate a true maker atmosphere in the library, students have to feel comfortable collaborating with each other. Collaboration can be noisy but there is nothing more exciting than hearing the buzz of collaboration between students. Of course, there still must be rules, but loosen up on the rule about no talking. Students will begin to see the library as a collaborative space, and this will benefit not only student learning, but also how the school library is seen by others. When teachers and administrators see and hear the excited energy created by students collaborating, the value of the school library will be affirmed again and again.

Promote and Market What the School Library Offers

As 3D design and printing catches on in the school library, make sure the word spreads about all of the great things going on in the library. Brief articles about what 3D printing is and how it is being used by students as part of the curriculum in a newsletter will pique the interest of parents and the community. Include photos of objects that students have 3D printed into staff and parent communications. If your school district accepts photos for the district website, send in photos of 3D printed objects, making sure to include the fact that the 3D printer is in the school library. Make sure that if the 3D printer was purchased in whole or partially by a parent-teacher organization or other group, members of the donating group are acknowledged and invited to attend a session so they can witness 3D printing in action.

Include Administration

Including campus and district administration is a must when beginning or maintaining a maker program, whether it includes 3D printing or not. Supportive administrators are vital to the success of any maker program. If administrators are unclear on the value of adding a 3D printer and maker area to the library, encourage them to follow principals, school librarians, and teachers on social media who have successful maker programs in their schools. Social media makes outreach and research very easy by connecting with educators who are already implementing and using new technology. Seeing real-life accounts of how the technology is being used and how it is benefiting students can be reassuring to administrators who aren't totally on board with it—yet.

Have a Maker Mentality

Teachers teach. That's what we do. But that's also what makes it difficult for school librarians and teachers to be the "guide on the side" when it comes to students learning 3D design and printing. It is tempting for librarians to teach a formal lesson on using the 3D design software and another formal lesson on how to use the 3D printer, if for no other reason than for the sake of time. Our time with students is limited because there are so many demands for time, and rightly so.

While difficult, it can be more effective for the school librarian to provide guidance for students during the 3D design and printing process without providing long how-to lessons. Allowing time for students to play with the software and learn the 3D design and printing process will be much more effective than feeding it to them. If time is limited, the school librarian or students could create guides, whether digital or on paper, for students to refer to while learning the software or if they run into a problem.

One of the best ways to ensure students are successful with 3D design and printing is to give them room to make decisions about their design while providing guidance and asking good questions.

Take Advantage of Students' Love of Video Games . . . and Anything Else They Love

Because many students play video games that are 3D based, it can be much easier for them to create and manipulate objects in 3D design software. Allow students time to develop objects related to their individual interests, including video games. Students who aren't that excited about 3D design and printing may get the spark when they realize they can create their own Minecraft tools in a 3D design software program and then upload them into their Minecraft game.

Practical Tips for Less Stress and More Fun!

1. Don't Panic When the Printer Is in Constant Use

There will come a time when there is a backup of objects waiting to be printed. As students begin to understand the possibilities of 3D printing and see what other students are printing, interest in 3D design and printing will soar. Some school librarians have reported having to purchase additional 3D printers to help meet the demand of students with a desire to print objects.

Develop a system for the order in which objects are printed and that will help when you feel like demand for printing is growing exponentially. Students can complete a simple form or sign up on a list on which they indicate their name, student identification number, a description of the object they wish to print, and possibly their preference of filament color.

Name	Student ID	Name of Saved Object	Filament Color Preference

The school librarian would then be able to let students know when their turn to print has arrived by looking up a specific student's schedule or by having students check back to see where they are on the list. If students have a preference of filament color they wish to use to print, the librarian could use the form or list to schedule printing by students requesting the same color of filament. This would help prevent the wasting of filament that occurs when a new color is loaded.

Some academic and public libraries have begun to charge students for 3D printing, and this is a viable option for cash-strapped school libraries for 3D printing that students do outside of classroom projects. If the school librarian wishes to pursue this option, ensure that administrators support the policy and that there are options for students who cannot afford to pay the cost of printing. Usual charges are somewhere around $0.05 per gram of filament used, which makes most medium-sized printed objects less than $5 to print.

Sometimes, two or more teachers might want to bring their classes to the library for a 3D modeling and printing project during the same period. The librarian will need to consider this possibility and communicate the guidelines for this to teachers. The librarian can have an online calendar where teachers can sign up for a project time so they will be able to see if other teachers have signed up for that same time period. If the librarian has volunteers that are willing to help out with 3D modeling and printing, more printing can occur during a time period.

2. Create Specific Policies for 3D Printing

In addition to library policies regarding checkout of materials, it may be helpful to have specific policies regarding 3D printing. A school library's 3D printing policy might include the following:

- The 3D printer may be used only for lawful purposes and cannot be used to create objects that are:
  - prohibited by local, state, or federal law,
  - harmful, dangerous, or that pose an immediate threat to the well-being of others,
  - obscene or otherwise inappropriate for the school library environment, or
  - subject to copyright, patent or trademark protection.
- The school librarian and library staff reserve the right to refuse any 3D print request.
- The current cost of printing an object
- The procedure for pickup of printed objects. For example, whether the object must be picked up by the individual who printed it.
- Rules for using the 3D printer, such as who can operate the printer and change filament.

It may also be helpful to post both policies and procedures for printing, making them widely available to students, parents, and staff. Below is an example of procedures that can be used to customize procedures for school libraries. Each library is unique, so policies and procedures will be unique. Librarians should create policies and procedures that work for their situation and library. Librarians can pick and choose policies and procedures and even add their own. The important thing is that the policy or procedure covers what it needs to for each individual library.

Example:

PROCEDURES

Design creation:

The 3D printer can be used with basic knowledge of 3D design software. Programs students are familiar with are __________ and __________.

Any 3D drafting software may be used to create a design as long as the file can be saved in .stl, .obj, or .thing file format.

The school library has computers with software that may be used to create a design.

Digital designs also are available from file-sharing databases such as Thingiverse.com.

Submitting a design for printing:

Students wanting to use the 3D printer will bring their file in .stl, .obj, or .thing file format (no larger than __ MB) to the school librarian.

Each student will be limited to printing one object per week, unless the object is part of a school project.

School library staff will preview the file and, if approved, prepare the file for printing.

The student will complete the required form and include all needed information.

Students will be notified when the object has been printed. Objects may be picked up at the library by the student who made the print request. Students must show a current student identification card to be able to pick up a printed .object.

3. See Failed Prints as Lessons Learned

Expect that students will print some objects that they are not happy with because the design wasn't as good as it could have been. Providing a system for designs to be double and triple checked prior to printing will help ensure the number of failed prints is lower, but they *will* still happen. Try not to see the failed print only as wasted filament, and instead, use it to get students to think about why the design failed. Because students will have saved the design in the program, using their name and student identification number to distinguish their design, they will be able to reopen the design, make improvements, and reprint.

If students are made to feel that a failed print is a terrible thing, they may lose interest in 3D design and printing and be afraid to try again for fear they will have a second failed print. An example of a great way to handle failed prints is at The DeLaMare Science and Engineering Library at the University of Nevada at Reno, where a "Box of Disappointment" is displayed in which students place printed objects that didn't turn out quite as the student intended. Keeping and displaying objects that aren't perfect shows students that failure is not the end of the world and that everyone encounters it at some point in the process.

That being said, misprints *do* waste time and materials, both of which are at a premium in a school setting. 3D printing companies are recognizing this and are creating programs to correct models before they are printed. For example, a program designed at the Free University in Berlin, called trinckle 3D (www.trinckle.com), fixes model designs before printing to help avoid misprints. Models are uploaded into the free, browser-based program and algorithms in CAD files are used to detect errors in the 3D model. Once an error is detected, trinckle 3D corrects it automatically.

4. Things to Be Aware of When 3D Printing

Printing Time

One of the primary downsides of 3D printing is the length of time it takes to print an object. Print time depends on the printer and the complexity of the design of the object, but objects can take 1–3 hours to print. Larger or more complex objects can take even longer to print. Even the 3D design stage can take several days, depending on how much time students are able to spend in the library and their experience level with the design software. Because of scheduling, it can be difficult for students to see the beginning and ending of a print job. Only smaller, less-complex objects would be able to be printed in the span of a library visit.

The school librarian can work with the classroom teacher to allow students to visit the library to see objects they have designed being printed. Because students save their designed objects using their names and identification numbers, the librarian can easily let students and their classroom teacher know when that particular object is being printed. Another way to share the printing process with students is for someone to video the librarian loading the filament, starting the printing, and skipping to the last few minutes of printing if you don't want the entire print to be filmed. The librarian can post these videos to a blog or website so students can access them anytime they want and can see the object they designed actually being printed.

The school librarian can utilize the time before and after school and student lunch times to 3D print student designs. This can not only help promote the school library but can provide students with learning opportunities during lunch. Opening up the entire makerspace to students during lunch will provide more traffic for the library than ever before. It may begin slowly, but as word gets around with the maker opportunities in the library, it will be standing room only. This option can work with secondary or elementary students.

The school librarian will want to enlist help from parents or staff to provide adequate supervision during lunch times as the crowd grows. Some students will not need much help, while others may need much more help in getting started.

3D modeling and printing can be integrated into all curriculum areas

Incorporating 3D modeling and printing into all curriculum areas benefits students by providing exciting and engaging learning experiences. Adding 3D printing as a tool will immerse students in the engineering process across the curriculum. When 3D design and printing are truly integrated into the curriculum, students become excited about school and learning. Utilizing 3D modeling and printing as an instructional tool provides librarians with an additional way to present information to facilitate learning for students and helps students create their own learning. By asking effective framing questions, students are encouraged to create meaning for themselves, which is when deep learning takes place. Many students crave more of a challenge in education, and 3D modeling and printing can provide that. Other students are challenged in math or reading, but have lost interest in school because they aren't engaged enough. 3D modeling and printing can solve this problem, as well.

Be safe and teach students to be safe, too

Because 3D printing is relatively new to the education scene, there is not much research available about the safety of 3D printing and children. The following are a few tips to keep in mind:

- Ensure that 3D printing takes place in a well-ventilated area. Although PLA filament has been found to be nontoxic, it pays to be cautious because we are working with children. Ensure the area is large enough so there are no problems with fumes. Ultrafine particles are given off by many pieces of equipment, such as photocopiers, laser printers, cooking, and yes, 3D printing. Ultrafine particles are nanoscale particles that have been found to settle in the lungs when inhaled. A well-ventilated area can help prevent this.

- Avoid printing with ABS filament until more research is conducted on its safety. Research by Stephens et al. (2013) shows that breathing fumes exuded from ABS filament has produced adverse health effects in adults, so it's best to stay clear.
- Be cognizant of printing small pieces that young children could choke on. Even if not working directly with young children, remind older students to keep small pieces away from younger siblings.
- Turn the 3D printer off when not in use. For extra precaution, the printer can be unplugged as well.
- Don't leave the 3D printer unattended while an object is being printed. The extruder nozzle gets very hot and students could be tempted to touch it. Teach the rules and guidelines for observing the printer until students know the rules backwards and forwards.
- Teach students to keep fingers away from the extruder at all times. Teaching students to never touch certain parts of the 3D printer, even when it is turned off, will train them to interact with the printer safely.
- Teach students to watch out for long necklaces, lanyards, flowing sleeves, and long hair when around the 3D printer. Students are often taught science lab safety rules, and most of these rules apply to 3D printer use as well.

Chapter 14 Key Points

Add a 3D printer to an existing makerspace in the library.

Include teachers, staff, parents, and community members in the 3D modeling and printing process. Connect with parents who may be already using 3D modeling and printing in their careers so they can help teach students and staff.

Open the library during open house or other schoolwide events and allow students and their parents to design an object that can be 3D printed. The librarian can print the objects and let students know when to come pick them up.

Including brief articles about how 3D printing is being used by students as part of the curriculum in a newsletter will pique the interest of parents and the community.

Embrace the maker mentality.

Create policies specific to 3D printing.

Safety first! Teach students how to be safe while using the 3D printer and hold them to it.

Glossary

3D Modeling – The creation of a digital representation of an object that has a length, width, and depth.

3D Printing – The process for making a physical object from a three-dimensional digital model, typically by laying down many successive thin layers of a material.

3D Printing Pen – A portable version of a 3D printer. Users draw in the air or on surfaces while filament extrudes to build an object.

ABS (Acrylonitrile butadiene styrene) Filament – Filament that is petroleum based and comprised of acrylonitrile, butadiene, and styrene. Not recommended for school settings due to possible toxicity when heated.

Additive Technology Manufacturing – Objects are built by adding one layer of material at a time.

.AVI file – Audio Video Interleave file. This is a common file format for audio and video data on a PC.

Axis – A fixed reference line for the measurement of coordinates.

.BMP file – Bitmap image file. File format used for pixel-based images.

Cartesian Coordinate System – a coordinate system that specifies each point uniquely in a plane by a pair of numerical coordinates for the *X*, *Y*, and *Z* axes.

Crowdfunding – The practice of funding a project by raising many small amounts of money from a large number of people, typically via the Internet.

Extruder – Part that handles the feeding and extruding of the build material.

Filament – Thin strands of material, usually plastic, used for 3D printing. See also ABS Filament and PLA Filament.

G-Code – G programming language used in computer-aided engineering. This is the language used to tell a 3D printer how to print an object.

.JPG (or JPEG) file – A common file format for digital photos and graphics.

Leveling – The process of setting the print bed the same distance apart from all points on the *X* and *Y* axes.

Open Source - A program in which the source code is available to the general public for use and/or modification from its original design free of charge.

PLA (Polylactic Acid) Filament – Filament made from plant-based materials. PLA can be made from sugar cane, beet cane, corn, or other vegetable waste and is biodegradable.

Print Bed – The type of surface the object is printed on and whether it is a heated surface or not. See also Print Plate and Print Surface.

Print Plate – The type of surface the object is printed on and whether it is a heated surface or not. See also Print Bed and Print Surface.

Print Speed – The speed at which an object is printed.

Print Surface – The type of surface the object is printed on and whether it is a heated surface or not. See also Print Bed and Print Plate.

Print Volume – Refers to the largest size at which an object can be printed.

Raft – A woven layer of sparse layers of filament that provide a base for the object being printed.

.STL file format – The standard format used for 3D printing and contains a list of the *X*, *Y*, and *Z* coordinates of the vertices for the object being printed.

WebGL – Web Graphics Library is a library of images that allows users to create and manipulate 3D objects without using plugins in a browser.

X Axis – Specifies the width of a 3D model.

Y Axis – Specifies the height of a 3D model.

Z Axis – Specifies the depth of a 3D model.

Resources

3D Modeling Libraries

3Dhacker.com
Cgtrader.com/3d-print-models
Cubehero.com
Fabster.com/physibles
Grabcad.com/library
Repables.com
Thingiverse.com
Youmagine.com

Books and Magazines

3D Modeling and Printing with Tinkercad: Create and Print Your Own 3D Models by James Floyd Kelly.

3D Printing: The Next Industrial Revolution by Christopher Barnatt.

3D Printing with Autodesk 123D, Tinkercad, and MakerBot by Lydia Cline.

Fabricated: The New World of 3D Printing by Hod Lipson and Melba Kurman.

Getting Started with MakerBot by Bre Pettis, Anna Kaziunas France, and Jay Shergill.

Make (magazine) – makezine.com. *Make* is available for digital and/or print subscription and contains many ideas for all kinds of making, not just 3D printing. Worth the price alone for the annual 3D Printing Guide.

The Book on 3D Printing by Isaac Budmen.

The Invent To Learn Guide to 3D Printing in the Classroom: Recipes for Success by David Thornburg, Norma Thornburg, and Sara Armstrong.

3D Printer Manufacturers

Cubify.com
Makerbot.com
Makergeeks.com
Printrbot.com
Stratasys.com
Reprap.org

3D Design Software

123D Design – www.123dapp.com/design
3D Crafter – amabilis.com
3D Tin – 3DTin.com
Anim8or – Anim8or.com
Blender – Blender.org
freeCAD – ar-cad.org
Inkscape – Inkscape.org
K-3D – K-3D.org
LEGO Digital Designer – ldd.lego.com/en-us
Nowmakethis – Nowmakethis.com
OpenSCAD – Openscad.org
Printcraft – Printcraft.org
Sculptris – Pixologic.com/sculptris
Seamless 3D – Seamless3D.com
Sketchup – Sketchup.com
Shapesmith – Shapesmith.net
Sweet Home 3D – Sweethome3D.com
Tinkercad – Tinkercad.com

3D Design Apps

123D Catch (iTunes)
123D Creature (iTunes)
123D Design (iTunes)
123D Sculpt (iTunes)
Artist 3D – Modeling Tool (iTunes)
Blokify (iTunes)
CreatureShow (iTunes)
Cubify Draw (iTunes)
Let's Create! Pottery (iTunes and Google Play)
Makers Empire (iTunes and Google Play)
MeCube (iTunes and Google Play)
Modio (iTunes)
Monstermatic (iTunes)
SpaceDraw (Google Play)
Verto Studio (iTunes)

3D Printing Pen Manufacturers

3Doodler – The3doodler.com
Creopop – Creopop.com
Lix – Lixpen.com

Websites and Digital Resources

3dprintingforbeginners.com
DesignMakeTeach.com – ideas and tutorials for including 3D printing in schools
EIE.org – Engineering is Elementary website
Makered.org – a great resource library for makerspaces for children
Makershed.com – the online store for Make:
Makerspace.com – the online community for Make:

http://3d.si.edu/ – Smithsonian Institute Digitization Project
Instructables.com – resources for many areas of making. To find 3D specific ideas, just search for "3D printing".

Filament Sources

Usually, compatibility with specific printers is specifically stated, but if not, call to make sure before ordering.
102creations.com
3ddirect.com
3dinkspot.com
3dprinterhub.com
Amazon.com
Fairwagon.com
GeckoTek print bed – http://www.geckotek3d.com/
Makerbot.com/filament
Makerlibre.com
Toybuilderlabs.com

References

American Association of School Librarians. 2007. *Standards for the 21st-Century Learner*. Chicago: American Association of School Librarians.

Brookhart, S. 2013. "Assessing Creativity." *Educational Leadership* 70 (5): 28–34

Churches, Andrew. 2009. *Bloom's Taxonomy: Introduction*. Educational origami. Retrieved September 9, 2014, from http://edorigami.wikispaces.com/Bloom%27s+-+Introduction

Collins, Owen. 2012. "Affordable 3D Printing." *Theater Design and Technology* 48 (1): 10–19.

Ennis, R. J. 1989. "Critical Thinking and Subject Specificity." *Educational Researcher* 18 (3): 4–10.

Grover, S. 2012. "How 3D Printers Work." Graphic. Licensed under Creative Commons Attribution license 3.0. Retrieved from http://www.thingiverse.com/thing:29432

Heater, B. 2014. *Manufacturing the Future: How 3D Printing Went from Pipe Dream to Your Desktop*. Retrieved from http://www.digitaltrends.com/features/manufacturing-future-strange-past-impossible-future-3d-printing/

Hedges, L. E. 1991. *Helping Students Develop Thinking Skills through the Problem-Solving Approach to Teaching*. Columbus, OH: The Ohio State University.

International Society for Technology in Education. 2007. *National Educational Technology Standards for Students*. Retrieved from http://www.iste.org/standards/standards-for-students.

Johnson, L., S. Adams Becker, V. Estrada, and S. Martín. 2013. *Technology Outlook for STEM+ Education 2013–2018: An NMC Horizon Project Sector Analysis*. Austin, TX: The New Media Consortium.

National Governors Association Center for Best Practices & Council of Chief State School Officers. 2010. *Common Core State Standards*. Washington, DC: Authors.

NGSS Lead States. 2013. *Next Generation Science Standards: For States, By States*. Washington, DC: The National Academies Press.

Norris, S. P., and R. H. Ennis. 1989. *Evaluating Critical Thinking*. Pacific Grove, CA: Midwest Publications.

Paul, R., and L. Elder. 2013. *30 Days to Better Thinking and Better Living through Critical Thinking: A Guide for Improving Every Aspect of Your Life*. Upper Saddle River, NJ: FT Press.

Prensky, M. 2008. "Turning on the Lights." *Educational Leadership* 65 (6): 40–45.

Schaper, M., R. Thompson, and K. Detwiler-Okabayashi. 1994. "Respiratory Responses of Mice Exposed to Thermal Decomposition Products from Polymers Heated at and above Workplace Processing Temperatures." *American Industrial Hygiene Association Journal* 55: 924–934.

Stephens, B., P. Azimi, Z. El Orch, and T. Ramos. 2013. "Ultrafine Particle Emissions from Desktop 3D Printers." *Atmospheric Environment* 79: 334–339.

3DTin Blog. 2014. "2D to 3D." Web blog post. June 14, 2014.

Wai, J., D. Lubinski, and C. Benbow. 2009. "Spatial Ability for STEM Domains: Aligning over 50 Years of Cumulative Psychological Knowledge Solidifies Its Importance." *Journal of Educational Psychology* 101 (4): 817–835.

Index

About the Author

LESLEY M. CANO is an elementary school librarian in Arlington, Texas. She has been a K–12 educational diagnostician and adjunct professor, and she has presented in Texas on technology and makerspaces in libraries. She has received over $50,000 in grants for her library and makerspace. Lesley was named the General Motors Arlington 2015 STEM Innovative Educator of the Year.